EFFECTIVE FUNCTIONAL VERIFICATION

Effective Functional Verification

Principles and Processes

by

SRIVATSA VASUDEVAN

A C.I.P. Catalogue record for this book is available from the Library of Congress.

ISBN-10 1-4614-9836-8 (HB)
ISBN-13 978-1-4614-9836-0 (HB)
ISBN-10 0-387-32620-0 (eBook)
ISBN-13 978-0-387-32620-7 (eBook)

Published by Springer,
P.O. Box 17, 3300 AA Dordrecht, The Netherlands.

www.springer.com

Printed on acid-free paper

This book is dedicated to:

The Almighty, who gave me a variety of experiences
and the inspiration to write this work.

My family for putting up with me as I wrote this book.

My colleagues who taught me what they knew
and worked with me.

My managers and mentors who encouraged me
on this incredible journey.

Contents

Part IV Ten Steps to Success

List of Figures

Foreword

Verification – both pre-silicon and post-silicon has become the dominant activity in all new semiconductor designs. There are many parallels to it in the software world – it is in the specification and initial development that determines the level of quality and reliability of software. Software Quality Assurance and Software Engineering has rigorous standards of continuous improvement of quality measurements that are used to track and continuously improve the systematic improvement of verification of software, the controlling of the interfaces, the planning of the unit level and module level tests, the bring up and maintenance of these modules, and the test and regression and isolation and workarounds involves with the software.

Digital design complexity is constantly increasing, and yet the thought put into these architectures with the hundreds of billions and trillions of transistors used has only recently become focused on predictability and techniques to drive quality throughout the RTL that is the dominant focus. In fact the most interesting point of research and activity today in building processors and SOCs.

This book is dedicated to verification exclusively. For something that is becoming the dominant focus of all designs – ie there are many books on architecture, modeling, designing for performance, for low power, for complex systems, but few to the real heart of the matter – verifying the digital design itself.

What you will learn here will influence the types of designs and the complexity of the design and your success in making these a reality. Ignore verification at your peril, embrace it in every aspect of planning and executing a project and your chances of succeeding are of course never certain, but much more likely to complete and complete to plan.

Srivatsa has been a driver in these techniques at Texas Instruments, and the many techniques described here are deployed, about to be deployed, or we will insist will be deployed here and at other companies. There are both case studies, state of the art techniques, and possible directions of studies the industry is going to move to. Verification is not an exact science, there is always the saying in the industry that it is an infinitely long task gated by a hard and fast milestone that couldn't be moved – ie a Tapeout date. Here is hoping that someday we can chance that.

David Witt
Director, World Wide Wireless Digital Design
Wireless Terminal Business Unit
Texas Instruments Inc.

Preface

Introduction

The world of semiconductor IC design has changed dramatically over the last few years. In the earlier days, IC designs were confined to a section of a building. Today IC designs are being created by bringing together diverse talent in multiple time zones in multiple continents across the world. Some companies outsource portions of verification as well. Interestingly some of these outsourcing activities are done in other countries to add to it all! No doubt, we have all made life interesting in the Silicon Age!

Why did I write this book?

As fortune would have it, I have been involved with ASIC verification for almost my entire career. I have had the good fortune of working with some of the best and brightest hardworking minds in Silicon Valley, CA, USA. During the course of my career, I have had the opportunity to be exposed to a variety of different verification environments and different types of IC designs. As I started working with young engineers whom I was mentoring, I realized that I needed a different set of tools to help the engineers and myself make our jobs easier.

Many engineers would come up to me and ask me for a "brain dump" on how to do verification. I also needed a platform to share concepts of verification with some of the junior engineers as well. I wrote this book to share experiences and a collection of best practices which have served me well over the years.

What is this book all about?

This is a book on the various practical aspects of ASIC functional verification. As chip designs get more and more complex, new tools are evolving to meet the challenges of functional verification. At the same time, many new engineers are joining this challenging field. This book addresses a variety of topics in this growing field.

Verification is no doubt a vast field, hence, I have chosen to focus a little bit more on the various principles in functional verification rather than the actual act of verification and test-benches since there are many excellent books on the subject. This approach is chosen since I believe verification is methodology and execution combined.

While discussing verification, I have not chosen to discuss any tool specific issues or methods since these are covered quite extensively in other texts. The material presented is tool/language agnostic and attempts to impart concepts and principles to make verification effective. The reader is encouraged to use the ideas presented herein in an implementation of the readers choosing.

Methodology and execution are extremely important in modern ASIC designs. Hence, winning strategies for verification are discussed along with ideas that improve the odds of first time success are made the focus of this book.

This book is organized into four major parts. There are multiple chapters in each part dealing with the theme of that part. The title pages in each part contain details of each of the chapters in that part.

Starting the verification journey: Part 1 discusses verification from an introductory point of view. Information in this part helps set the baseline for the other chapters in the book. Engineers and managers new to verification or considering changing over to verification as a career would find this section informative.

Ingredients of successful verification: Part 2 begins by describing the human aspects of verification along with case studies from the real world. Metrics that drive verification are presented here as well. Managers and lead engineers might find information in this section more useful. It is hoped that the information in part two also helps to motivate newcomers in the field.

Reducing work in verification: Part 3 discusses various methods to reduce time, effort, energy and money during the course of verification. Various methods and ideas are presented that can be adopted by the reader in their own verification environments. While this section is partly targeted at the advanced user, it is hoped that the reader finds inspiration and motivations from the chapter presented in this section. The concepts presented in this section are designed to help engineers at all experience levels become productive in verification quickly.

Bringing it all together: Part 4 presents a process and methodology, which attempts to bring the entire verification effort. The principles in Part 4 have been time tested in a variety of organizations. These principles help ensure a smoother verification "experience" as well as a successful tapeout on silicon. It goes ahead to describe 10 steps that would be helpful to the engineer to get from the 'concept' point to the actual tapeout of the device.

Who should read this book?

I would really like to say everybody who makes a career of ASIC development! Actually, the book itself is written with a wide audience in mind. My hope is that this book will inspire many more people to take on the challenge of verification and finds use in the hands of all engineers as a reference.

How should this book be used?

Simple! Read it from cover to cover. Implement some of the concepts presented herein if you find them useful. Please do share and discuss the information with your friends to implement useful ideas and do not let this book rot in your bookcase! Whatever you do, please do not use it as a doorstop or a paperweight, or something similar!

Honestly, there really is no best way. The book itself is presented as a two column format with the essence of the discussion presented on the left hand column and a discussion of the topic on the right. I invite readers to pick the chapters of interest to them and dive right in. Section 1 and 2 are for everyone. I do suggest that you, the reader, look into all chapters so that you get an idea of the subject matter along with a perspective of the author. The index at the end of this book along with the table of contents should help anyone find information in this book quickly.

There is no doubt a lot of literature on hardware verification languages and methodologies. Hence, it is hoped that this book be used as an advanced textbook to teach new engineers about ASIC functional verification while supplementing presentations in the classroom as a reference. Engineers who have just begun ASIC verification as a career can use the information in this book to quickly ramp up their skills.

It is also hoped that the readers find inspiration and use some of the concepts presented herein. There are certainly many other extremely valid and differing viewpoints on various aspects of verification presented in this book. I would be grateful to hear of them so that we may all enrich one another. My contact information is enclosed below.

Contacting the Author:

I suppose nothing could be easier than typing up an email to srivatsa@effective-verification.com. I would appreciate any communication that would help improve the contents of the book and will gratefully publish it along with the errata for the book on http://www.effective-verification.com.

Acknowledgments

I thank the Grace of the Almighty ONE without whom none of the experiences would have ever happened to me. Much inspiration and strength has been derived from HIM. It's been interesting journey so far.

My parents have been a source of strength. I am especially indebted to my family members who were exceedingly patient with me as I worked on this book. In particular, my wife has been a constant source of inspiration when the work seemed large and helped me get through many difficult spots all along the way.

Janick Bergeron of Synopsys and Steve Dondershine of Texas Instruments reviewed this book and gave me detailed feedback. My deepest gratitude to them for taking valuable time out of their busy schedules to review this book.

I am also very grateful to all my colleagues who had shared with me their knowledge as I worked with them throughout my career. Many individuals helped me with the book. They have been named on the next page.

There are no doubt many other people who have not been mentioned who have contributed either indirectly or directly to this book. My deepest gratitude is offerred to all of them who helped me in my time of need.

Srivatsa Vasudevan
December 2005
Bangalore, India

The specific individuals named below amongst others helped me at various times to make this book a reality. My gratitude is offered to all of them.

Rakesh Cheerla	Extreme Networks
Vipin Verma	Texas Instruments Inc
Umesh Srikantiah	Infineon
Dr. Vijay Nagasamy	NEC Corporation
Mukul Tikotkar	Texas Instruments Inc
Ish Dham	Texas Instruments Inc
Nagaraj Subhramanyam	Texas Instruments Inc
David Witt	Texas Instruments Inc
Sanjay Palasamudram	Texas Instruments Inc
Bhaskar Karmakar	Texas Instruments Inc
Dr. Venu Gopinathan	Texas Instruments Inc
Dr. Mahesh Mehendale	Texas Instruments Inc
Venkata Rangavajjala	Tellabs Inc
Ashok Balivada.	Analog Devices Inc
Deborah Doherty	Springer Technical Support
Mark DeJongh and Cindy Zitter	Editors at Springer
Vidhu E.	The cover design for this book
Digvijay Lahe.	Fourth Dimension Inc
N. Raghunathan	Helping me edit the book
Shrinidhi Rao	Proofing the book
G. N Lakshmi	Figures in the book

PART I

INTRODUCTION TO VERIFICATION

This part provides a baseline for the rest of the chapters in this book. Information in this part is divided into the following three chapters:

Introduction to Verification: This chapter provides an introduction to verification. It describes the need for verification in contemporary designs. Various types of verification activities are discussed here along with an overview of the ASIC design process.

Approaches to Verification: There are many different ASIC designs in the marketplace. Each design merits a different approach. This section covers the various approaches to verification. The pro's and cons of various approaches are described in this section. The chapter also describes various levels of integration verification.

Verification Workflows : Different companies use different processes with varying degress of formalism to ensure quality of their designs. This chapter focusses on providing an overview of various workflow processes in verification. It offers the reader a perspective view of the verification processes that occur during the verification activity.

Chapter 1

AN INTRODUCTION TO IC VERIFICATION

The Need for Verification

Verification is the activity that determines the correctness of the design that is being created. It ensures that the design does meet the specifications required of the product and operates properly.

The free online dictionary of computing [1] defines verification as "*The process of determining whether or not the products of a given phase in the life-cycle fulfill a set of established requirements*".

In the IC design world, the process of ensuring that the design intent is mapped into its implementation correctly is termed as verification. Such a description is indeed broad. In the sections that follow, the verification activity is further divided into several major areas. These are introduced to the reader in this chapter.

1.1 Importance of Verification

Do we need to verify anything at all?

Ensuring that a product operates correctly as per the specifications is crucial to ensuring that the product will be used by its intended users.

In order to illustrate the importance of verification, consider a simple example of a digital camera being introduced by a company. A typical digital camera is made up of many components like the lens, the battery, the sensor, an image processor, software etc. All of these must work correctly before a picture is taken using the camera.

In this example, it is assumed that the camera design is conceived with an intention of capturing a large market share in its segment. In order to introduce this product into the market, the company's marketing team typically analyzes the products available in the marketplace and then creates a specification for a product that they believe will have a good impact on the market, bringing revenue to the company.

The requirements for the camera are analyzed by a systems team and a design for the camera which meets the requirements of the marketing team is created.

In this example, it is assumed that many of the parts for the camera are purchased from elsewhere. However, the core components of the camera are split into one or more ASIC devices that need to be created to satisfy the market requirement. The ASIC devices are then designed by ASIC designers. For a variety of reasons, it may not be possible for the designer to think of all the various possibilities and scenarios that the device may be used in when they are in the process of designing the device.

Consider a scenario where the product was actually sent out to the customer without making sure that the product actually worked, then there is the possibility that the company designing the product may suffer from significant financial losses, not to mention damage to the company's reputation! The author is confident that the reader would not like to buy a product if there was no assurance that the product actually worked!

Verification is that activity which ensures that the camera in the above example does indeed operate correctly as per the specifications. It ensures that the product that is created does not suffer from defects and operates satisfactorily.

Why is it so important?

Looking at a typical project team, one notes that in order to verify a design, one must first understand the specifications as well as the design and, more importantly, devise an entirely independent approach to the specification.

It is not unusual for a reasonably complex chip to go through multiple tape-outs before mass production in spite of a large investment of resources in verification. Verification to ensure correctness is therefore a must.

The effort of verification is greater than the actual design effort

A design engineer usually constructs the design based on cases representative of the specifications. The role of the design engineer ends with having translated the specification into an implementation that meets the architectural specifications. However, a verification engineer must verify the design under all cases, and there could be an apparently an almost infinite number of cases. It therefore follows that the amount of work required of the verification engineer to prove that the design actually works is much more than the design effort and must be treated as such.

From the makeup of a project team, one can see the importance of design verification. A typical project team usually has an equal number of design engineers and verification engineers. Sometimes design engineers are paired with a couple of verification engineers and hence even have a ratio of two to one.

Verification effort is different from the design effort

If the verification engineer follows the same design style or thought process as the design engineer, both would commit the same errors and little would be verified. Therefore, it is implied that the verification effort is an independent effort from the original design effort. The verification activity while being independent is considered a part of the overall design effort. There is the danger that a designer verifying his/her own block may end up verifying their *implentation* instead of the specification.

Verification is a costly business

Verification is unavoidable. It always costs too much and takes too long. However, proper verification techniques can save a company a significant amount of money [3]. Verification is now recognized as a very important activity by many companies.

Statistically, it has been shown that over 70% of an ASIC design budget is now devoted to verification [2]. Given that the typical ASIC device costs a few million dollars to develop and test, it becomes apparent that there is a lot of money spent in the verification activity. Hence, it is imperative that ASIC verification must be as effective as possible in order to be able to provide proper returns on investment.

History shows us the cost of a recall

Chip designs in recent years have become increasingly complex. In the recent past, because of incomplete verification, several companies had to recall their products and replace them with working ones. For example: Intel Corporation had to spend over 450 Million Dollars to replace faulty Pentium devices [2]. Another company, Transmeta, had initially several difficulties with their earlier parts as well [2].

1.2 Overview of a Typical ASIC Design Process

The earlier sections described the need and importance of verification in a product cycle. This section explores a typical ASIC design process and various aspects of verification that occur during this process. The entire process is shown in figure 1.1.

The ASIC design process begins with the definition of the various features of the system incorporating the device that will be designed. During this time, various inputs from market research, system feasibility, cost, profit margins etc. are determined.

The next stage in this process is the definition of the overall system architecture to address the product's needs. The architecture maps the product requirements into software and hardware components. The hardware component that arises is termed the hardware architecture of the system. A similar specification is derived for the software. The hardware component of the system is partitioned into one or more ASIC devices. This partitioning of the system functions helps define the various architectural features of each of the devices in the system.

The architectural specification of each of the devices is then translated into a functional specification for the device. During this phase the various structural elements of the device are then identified. The end result of this phase is the micro-architectural specification and functional specification for the device.

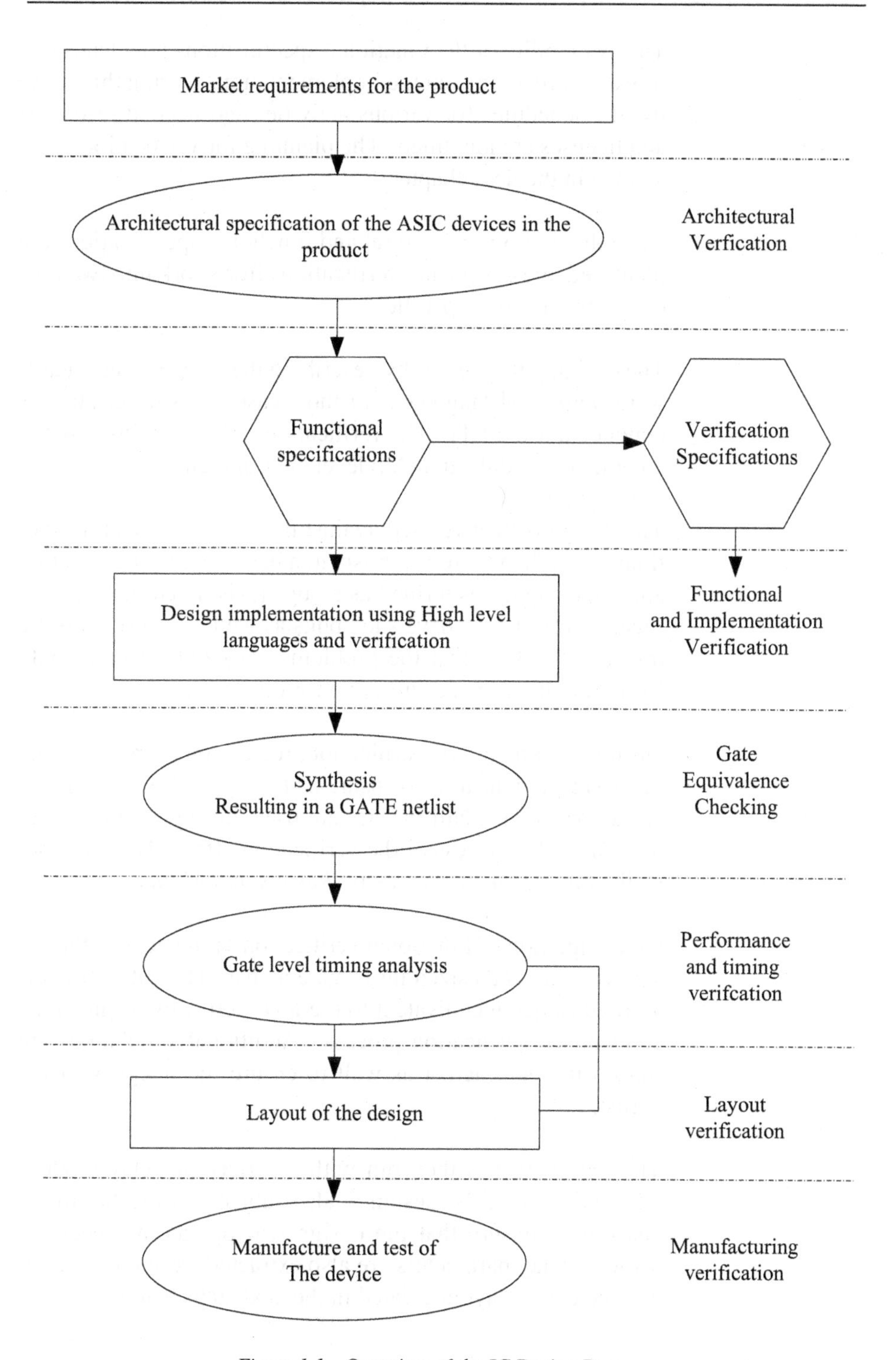

Figure 1.1. Overview of the IC Design Process

The availability of the functional specifications now leads to a process of planning out the implementation. During this activity, the schedules for various activities and needed resources and licenses are identified. The planning for verification is described in the next chapter.

Once the micro-architectural and functional specifications are identified, the design and verification efforts fork into two processes that occur in parallel.

The design effort begins by describing the device using a hardware high level language. In most cases, this description is conducted using RTL. The verification activity begins with the creation of module and chip level environments.

The design is then verified against the specifications for functionality. When there is a mismatch between the specification and the design, it is termed as a bug. (It is noted that the verification effort may have such bugs as well!). These bugs are tracked to ensure that the problems are fixed. This activity forms a bulk of the design cycle for the device.

During the process of verification, regressions[1] are run using the design and the tests to ensure that orderly progress is made. A number of standard metrics are used during this process to determine the success of the verification effort. These are described in the chapter *Tracking Results that Matter*.

On completion of functional verification of the device, the design is then synthesized into a gate netlist. The gate net list is verified as being equivalent to the RTL netlist by means of an equivalence verification process. Additional simulations are run on the gate netlist as well to ensure the integrity of the netlist.

The gate netlist is then run with a variety of delay models. At this juncture, the design is also subjected to static timing analysis to ensure that the design will operate as specified. Various delay parameters are also extracted from the libraries and the device layout created in the next steps below.

[1]Regressions are described in the chapter *Verification workflow processes*

The gate netlist is also used to generate various patterns that will be used to verify the device on the test equipment. These patterns help in generation of Functional Tester tests that are used to determine if the device that is fabricated is good or bad. These patterns are then used in the manufacturing verification process below.

Depending on the technology (standard cell/custom design) the gate netlist is then translated to a sequence of polygons in multiple layers to create a layout of the device. At this point, another verification process is deployed to ensure that the translation happened correctly between the gate netlist and the layout that has been generated[2].

The device is then fabricated. After this process, manufacturing verification usually involves making sure that the device was manufactured correctly and no flaws were introduced by the manufacturing process. This is accomplished using the patterns generated from the gate netlists.

From the description, it is apparent that verification occurs in a number of phases. During each verification phase, various activities are undertaken to ensure that the device is indeed successful in silicon. Some of the verification activities are well suited to the use of some automated tools which help make the task significantly easier.

Functional verification

Functional verification is the activity where the design or product is tested to make sure that all the functions of the device are indeed working as stated. This activity ensures that the features functions as specified. It is noted that some of the device features may or may not be visible to the user and may be internal to the design itself. However, it is imperative that all the features are verified to operate correctly as specified. This verification activity usually consumes the most time in the design cycle.

Gate Equivalence verification

Equivalence verification is the activity to ensure that the schematic generated or created is actually a true representation of the design specified and verified functionally above. At this stage

[2] This is called as LVS – Layout Vs Schematic

not much timing information is considered. Only logical and sequential relationships are considered.

Timing verification

Timing verification is the activity where the timing of the circuits is actually verified for various operating conditions after taking into account various parameters like temperature etc. This activity is usually done after all the functions of the feature are verified to be operating correctly. This is an important part of the verification activity.

Performance verification

Performance verification is the activity that could be carried out after an initial test device has been created as frequently happens with analog circuits or is carried on as a part of the functional verification activity. In some organizations, this term is also used for performance characterization of the device architecture.

Layout verification

Layout verification is an activity to ensure that the layout of the design indeed matches the schematic that was actually verified in the timing verification above.

Manufacturing verification

Manufacturing verification usually involves making sure that the device was manufactured correctly and no flaws were introduced as a result of the manufacturing process. During this process, the device that has been fabricated is tested using a tester which apply patterns to the device that has been fabricated. These patterns are determined using the gate simulations and ATPG tools.

The difference between testing and verification

One important question often asked is the difference between testing and verification. Verification is the activity that is used to verify that the design does indeed meet its specified intent. Testing on the other hand is used to ensure that the design is indeed manufactured properly. Testing is accomplished by the use of test patterns that are generated using a process termed as ATPG (Automatic Test Pattern Generation). Verification on the other hand involves the generation of test vectors which are used to ensure that the design meets the specification.

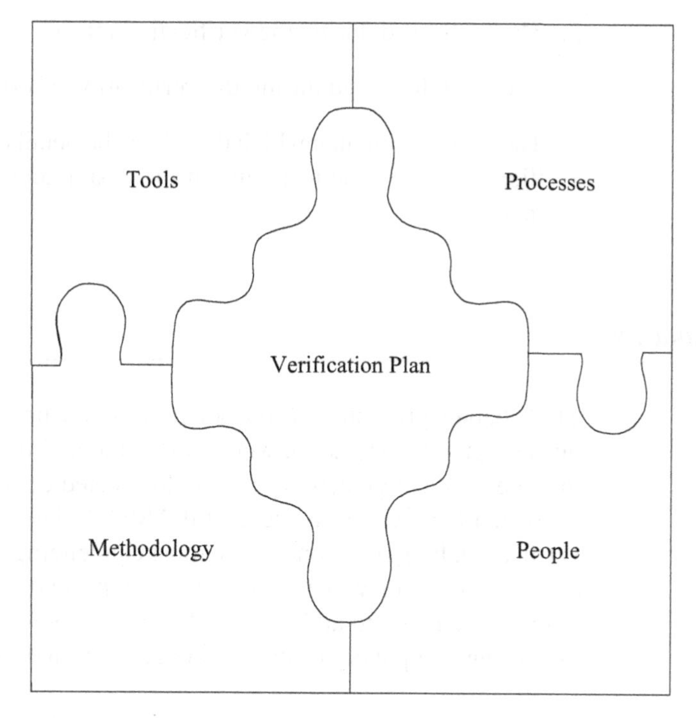

Figure 1.2. Factors in Successful Verification

1.3 Factors in Successful Verification

Various verification activities in the ASIC design flow have been described above. Each of these activities uses a multitude of different tools and techniques to achieve the goal of verification. They have also been researched extensively all over the world. Each of these areas merits several books to do justice to the discussion of the various topics involved!

Successful verification is a result of many ingredients coming together [4]. These ingredients are all part of a large jigsaw "puzzle" as can be seen from figure 1.2. Any successful ASIC design effort can be attributed to the following:

1. People factors which make an enormous impact on the project.

2. Methodology used to verify the device.

3. The tools used during the verification effort.
4. The workflow used during the verification effort.
5. The verification plan which describes the details of the verification effort, the tests run on the design and the results thereof.

Conclusions

This chapter presented an introduction to verification. Various aspects of verification were considered and it is apparent that the verification activity is a multi-faceted complex activity which is crucial to a successful ASIC design. Functional verification by large is one of the most challenging areas of IC design verification with significant research and product development occurring to address the challenges posed by the ever increasing complexity of the current generation of designs.

In the chapters that follow, various aspects of functional verification are discussed along with some examples.

References and Additional reading

[1] Dennis Howe. The free online dictionary of computing. 1993.

[2] Todd Austin. Building buggy chips that work! *Presentation from the Advanced Computer Lab*, 2001.

[3] Brian Bailey. The wake of the sleeping giant-verification. *www.mentor.com*, 2002.

[4] Brian Bailey. Verification strategies - the right strategy for you. *www.mentor.com*, 2002.

Chapter 2

APPROACHES TO VERIFICATION

Different Ways of Reaching the Goal

The previous chapter described the need for verification along with an exploration of the ASIC design process. A brief description of the various factors that affect the verification effort was also presented. In this chapter, basic verification principles as well as some common verification approaches are discussed.

The approach to the verification activity is just as important as the tools that are used to accomplish the job. Verification too is undergoing a process of rapid evolution. The tools that are used by the industry are improving at a rapid pace.

Complexity will only go up

Since the early days of the semiconductor industry, the complexity of the device under test has been steadily increasing. Along with the complexity, the risk of a single corner case bug that may cripple the company is also increasing. Various processes, tools, and techniques are used to counter this complexity challenge that is present in the industry today.

The cost of failure is rapidly becoming one of the most important issues in the semiconductor industry today. It has become apparent that no single tool, language, or approach can easily provide all the required features to create and manage a verification activity. The current verification environment today typically uses a multitude of scripts and languages to accomplish the task of verification. In addition, almost all verification environments today use some combination of approaches to achieve the goal of a successful tapeout. While there's no magic solution to the question of choosing one approach over another, it is noted that any art form is usually made up of a few principles and many techniques!

The various types of tools along with their description and operation can be easily found by the reader in a variety of literature. Hence, the author has chosen to concentrate on offering an introduction to various approaches to functional verification in this chapter. The pro's and con's of each approach are also presented in this chapter. In addition, references are provided to allow the reader to pursue in depth the approach of their choice. The approaches may also be combined as needed by the reader to suit the needs of the particular device under verification.

2.1 What is Functional Verification all About?

In the previous chapter it was observed that the design process transforms the specification of the device into an implementation as understood by the designers. On the other hand, verification is the process of ensuring that the transformation created by the designers is indeed an accurate representation of the specification. In Janick's book [18], he refers to this as a "reconvergence model" Such a process is shown diagrammatically in the figure below.

During the process of implementation, a human typically reads and understands the specification and then creates some RTL code that is then transformed into the design. During this transformation process, there is no doubt the possibility of misinterpretation or omissions of some aspects of the specification. It is not typically possible to take out the human factor as there are many things in this process which are not well defined.

The process of verification could be primarily accomplished by placing the device under test in a testbench and then applying some vectors to the design to ensure that the design does indeed meet the specification. The testbench takes over the task of applying inputs to the design and setting it up in a known configuration. Various input vectors are applied to the design to ensure that the response is as expected by means of tests that

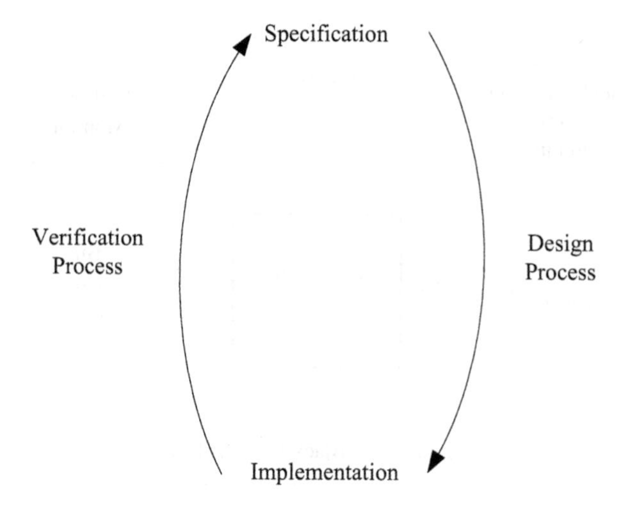

Figure 2.1. Verification versus Design

are run using the testbench. In addition to the device under test, various other modules which check the output of the device or observe some signals of the device under test may also be instantiated in the testbench. The checkers could perform various functions in the testbench. For instance, some checkers may check for a protocol on the inputs and outputs of the device. Monitors perform an additional function of watching the I/O or some specific busses in the device under test.

There are a variety of considerations when writing a testbench. These considerations are very well addressed in [18]. Some considerations on developing a testbench may also be found in the chapter *"Cutting the ties that bind"*.

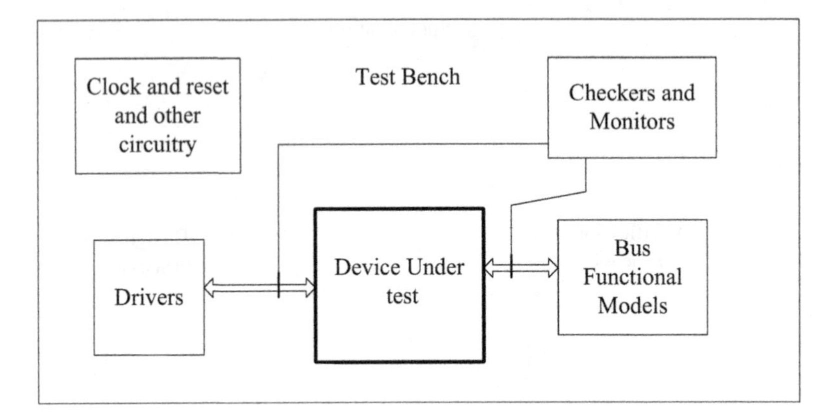

Figure 2.2. Black Box Verification

2.2 Stimulating the design - A choice of approaches

In the previous section, the basic concept of a testbench was addressed. The method of generating stimuli for the tests can be classified into three basic methods. These are:

- The black box method
- The white box method
- The gray box method

2.2.1 Black Box Approach

In this approach, the design to be tested is verified without having any knowledge of the details of the design. The verification activity is carried out using the external interfaces of the design. No information about the specific internal implementation is considered during testing from this approach. Stimuli are applied to the external inputs of the design and the response is observed on the output of the design. The pass and fail criteria for the design is determined by looking at the response for a certain input from the design and determining the correctness of the response for a certain input.

An example: In the figure 2.2, a simple SONET overhead processor is shown as a device under test in a testbench. The testbench is designed to drive the inputs of the device. The outputs of the device are collected by the testbench and processed to ensure that the device under test is actually able to implement the SONET frames correctly. Various error and disturbance conditions are injected into the device from the inputs and the response of the device is observed at the output of the device.

This approach suffers from an inherent limitation.

The black box approach is hampered by a significant lack of visibility and controllability. It is a large challenge to isolate where the problem is when a test fails since there is not much known about the design. It is only possible to determine that a specific test for the design either passed or failed. If there is a failure, one is hard pressed to determine exactly where the problem in the device is. It is also very difficult to use this approach to test some low level features buried deep in the design.

Black box verification techniques are not particularly well suited for doing block level implementation. The reason for this is somewhat obvious in that it becomes difficult to reach all of the interesting corner cases when nothing is known about the implementation methodology. If this information is known, it becomes much easier to write very specific tests to target the corner case behavior.

Conformance/standards testing use this approach.

Many a time, the design that is created may need to conform to certain standards. Under these conditions, a specific standard input is provided to the design and the response is observed. The design is deemed to have captured the intent of the specification if the design produces the appropriate output for the specified inputs.
Black box verification finds favor use when the design under test is created to conform to some specific industry standards (As indicated, a SONET processor as shown above, an Ethernet MAC) etc. Larger subsystems comprising of many ASICS also tend to use the black box approach to ensure that the design does indeed meet the intent of the specification.

Tests using a black box approach lend themselves to reuse

As can be observed, the black box approach does not contain any design specific information in the tests. As a result, it now becomes possible to reuse the tests (albeit with some modifications or abstractions) over a range of devices.

2.2.2 White Box Approach

The white box testing on the other hand implies that there is detailed knowledge of the design. There is knowledge of the internal workings, hierarchy, signals etc. As a result, the tests are written to test very low level details of the design. The figure 2.3 gives an example of such an approach.

In the figure, the details of the arbitration algorithm, the cache replacement algorithm and state machine are all exposed to the verification engineer. The engineer writes tests with full visibility of the design.

The tests developed make use of design knowledge

Detailed knowledge of the design makes it easier to ensure that various low level details of the design are indeed operating satisfactorily as indicated. In addition it is now possible to ensure that each and every component in the design is completely exercised. As a consequence, the verification engineer could help the design engineer find bugs more quickly.

Reuse is now more challenging with this approach

Since the tests make use of extensive design knowledge, reuse of the tests is a little more difficult since another design that attempts to reuse the same test may not have the same design features as the original design.

2.2.3 Gray Box Approach

The black box and the white box approaches both have some inherent advantages and disadvantages. The white box testing approach can then be combined with the black box approach so that a combination can be created with the advantages of both the white and black box approaches.

This middle ground is referred to as gray box verification. In this style we know about the general architecture of the solution. There is limited access to intermediate points,Which are usually inter block communication, and protocol compliance. There is ample opportunity to inject communications

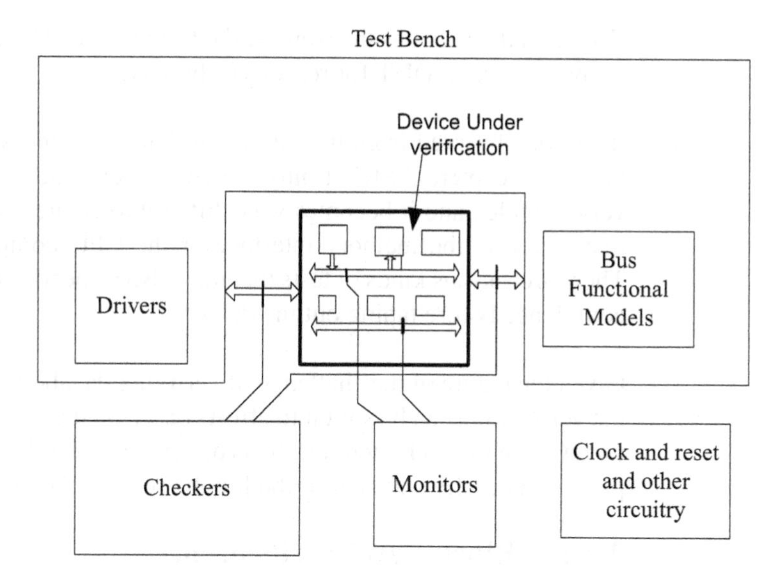

Figure 2.3. White and Gray Box testing approaches

between smaller blocks, standard protocols predominate. Dedicated monitors for these protocols can be inserted and in some cases may inject additional traffic in order to help verification throughput or congestion behaviors.

2.3 Verification Approaches Based on Integration

The earlier sections described various approaches to verify the design. In this section, various levels of integration are explored. As the reader will observe, with rising levels of integration, there is a corresponding rise in the level of abstraction in the test and the environment. This section is intended to help the reader visualize the use of various approaches in verifying a hypothetical device.

In the first chapter of this book, an example device of a camera ASIC was introduced. In the figure 2.4 a camera ASIC[1] is indicated. This device has many components including a camera interface, a MPEG module, a RAM memory interface, a memory card interface and a battery monitor interface along with other signal processing modules as shown in the figure 2.4.

[1] Similarity with the specifics of a company's product is purely coincidental and unintended!

(Note: In the sections that follow, the term Device Under Test is abbreviated to DUT for reasons of brevity.)

It can be observed from the figure 2.4 that it is impossible to verify the camera ASIC at only the chip level. The device is very complex and it becomes very difficult to create tests that may exercise the memory interface or the CPU completely. There are various kinds of tests that may also be needed to gain confidence before taping out the device.

Given the nature of the challenge of verifying the above ASIC, the verification can be split into efforts based on the level of integration of the RTL modules. This division follows the design process naturally and is described in the sections below.

2.3.1 Block Level Verification

The block level verification environment may contain a single module of RTL or several small modules grouped together. In the example indicated above, the camera interface, MPEG module, RAM memory interface, memory card interface, Battery monitor etc, are treated as a single blocks.

In this environment, the various inputs and outputs of the module are connected to bus functional models. These models mimic the behavior of the modules that would otherwise be connected to the DUT. Such an arrangement for the memory interface block indicated in figure 2.5.

In the block level environment, there is usually considerable flexibility with regards to the inputs and observability of the inputs/outputs of the DUT. The tests are a mixture of both black box and white box approaches and this approach is used to verify the device very thoroughly before integration into the larger subsystem blocks of the chip.

Different blocks in the above example have differing needs in verification. They may require different approaches. For example:

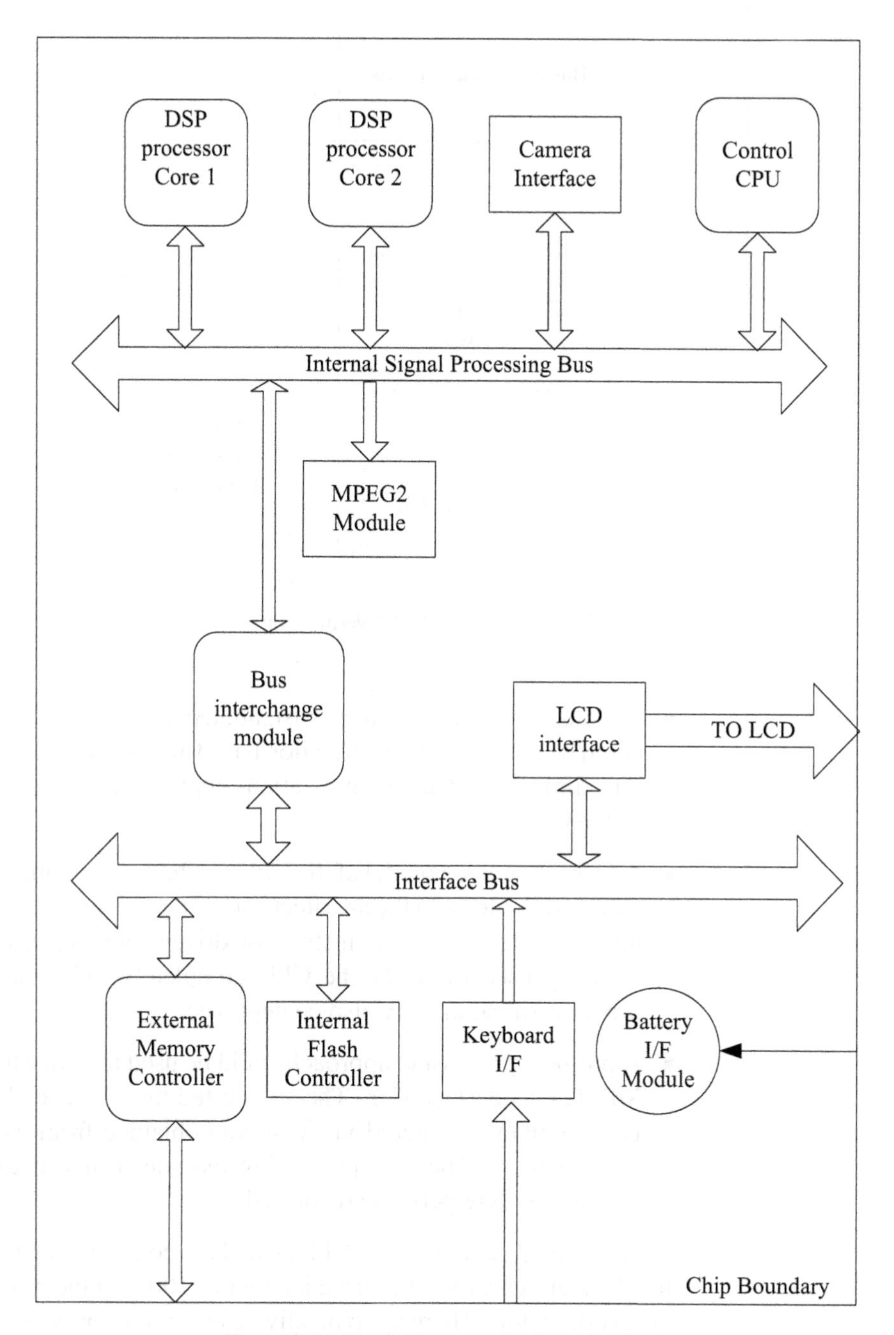

Figure 2.4. Block Diagram of a Complex SOC

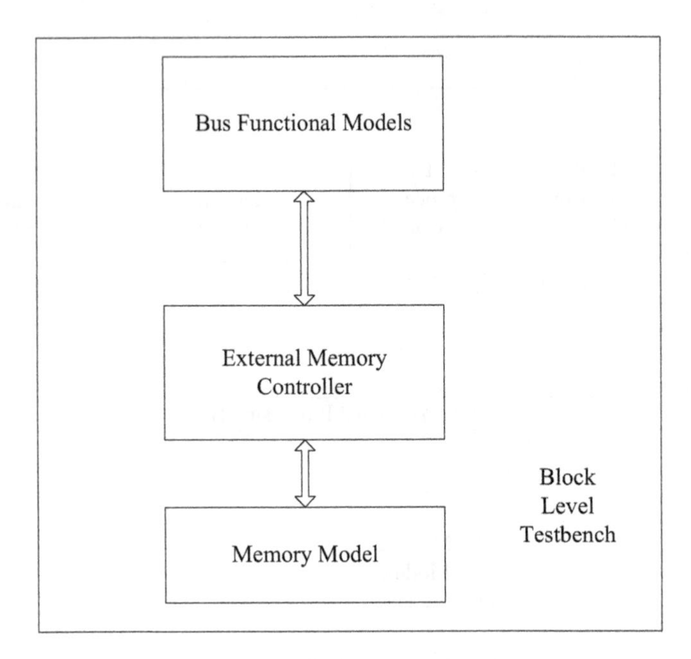

Figure 2.5. Block Level Verification

- Various memory interfaces instantiated in a block level could be tested using a transaction based approach with various timing and data randomizations applied during testing of the block.

- A software golden model of the CPU available is sometimes made available from the architects of the device, and a mixture of golden model and instruction driven approaches are typically used to verify the CPU completely. Coverage driven verification may also be deployed.

- A pre-post processing approach could be used to verify the MPEG encoder/decoder. The stream fed into the encoder is sometimes a standard video stream obtained from various sources. The post processing may determine if the operations were performed correctly.

The functional coverage of the block and the code coverage of the block are usually set to have a target of 100% in the block level verification. There are typically no exceptions or waivers granted for checking functionality in the block level.
One of the nice things about a block level environment is that there is usually significant control available to the verification

engineer. Various scenarios can be created with a ease, and the run time for the tests is usually very short. This allows for fast debug turnaround times. For the most part, many of the design bugs are caught in this phase.

2.3.2 Sub System Verification

In this level of integration, various subsystems are grouped together and verified for interactions between the blocks. Proper operation of the subsystem is also verified in this approach. The subsystem model is usually much larger than the block level environment. The bus functional models used to mimic the behavior of the RTL at the block level are now replaced by actual RTL in this level of integration. The run time for the simulations is typically longer than that of the block level verification. It is also more challenging to create very specific scenarios inside a block compared to the block level environment. Turn around times for bugs are larger in this level of integration. Many designer assumptions regarding the nature of the block interfaces (polarity, protocols, bus width etc) are usually flushed out in this level.

As seen in figure 2.6, the memory controller, the CPU and the camera subsystem are made a part of one of the subsystems. Other subsystems or groupings are also possible, but not indicated here. In some cases, some bus functional models (BFM's) are instantiated to mimic the behavior of another subsystem on the same device.

In this particular example, typical tests at this level could ensure that the CPU is able to program the memory controller and the camera block correctly and data from the camera block is indeed being stored in the RAM external to the device. Various randomizations on the data patterns, control patterns etc could be performed. The interactions between the various blocks and the inter-connectivity of the blocks could also be tested in this subsystem. Performance of the subsystem may also be verified.

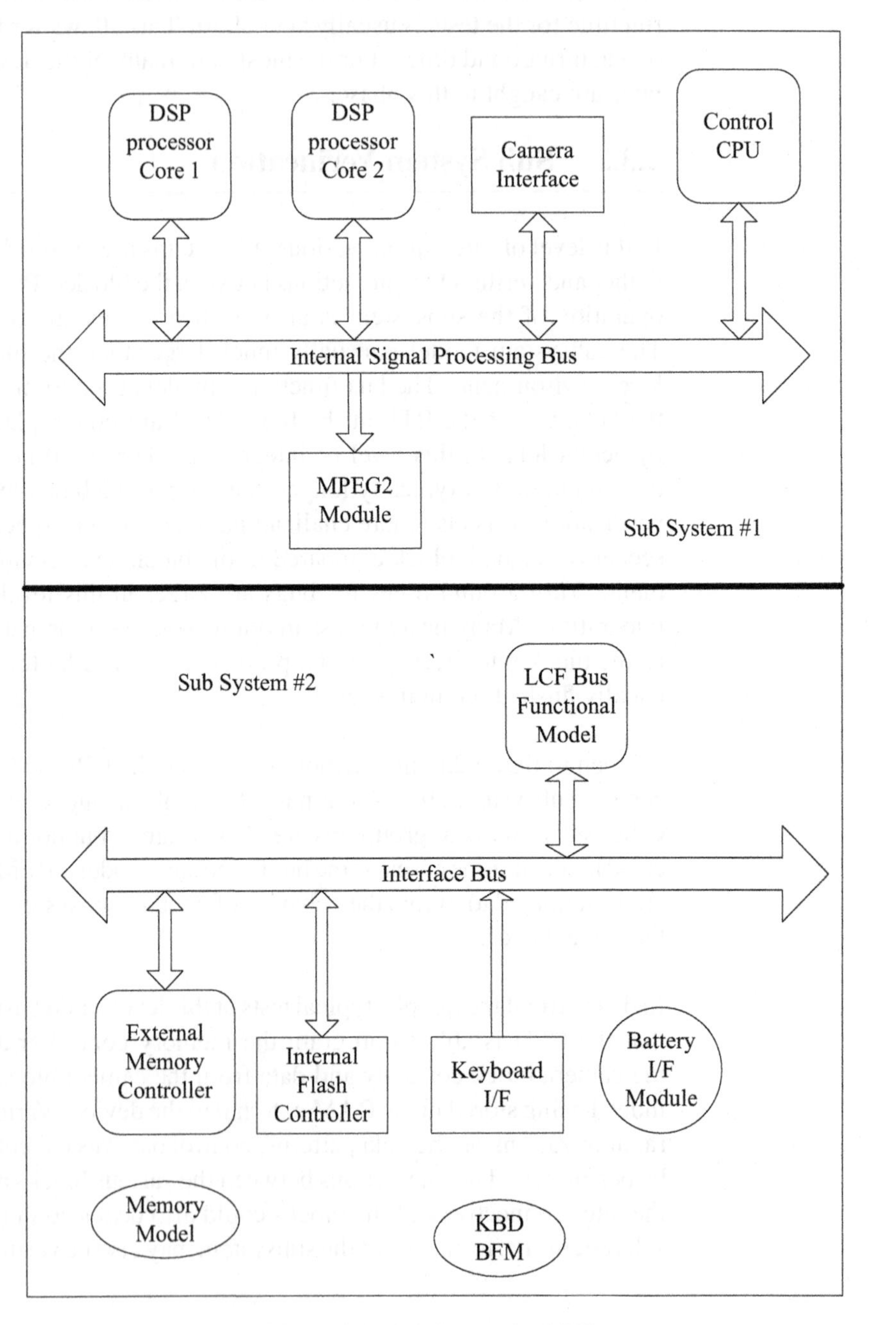

Figure 2.6. Sub System for ASIC

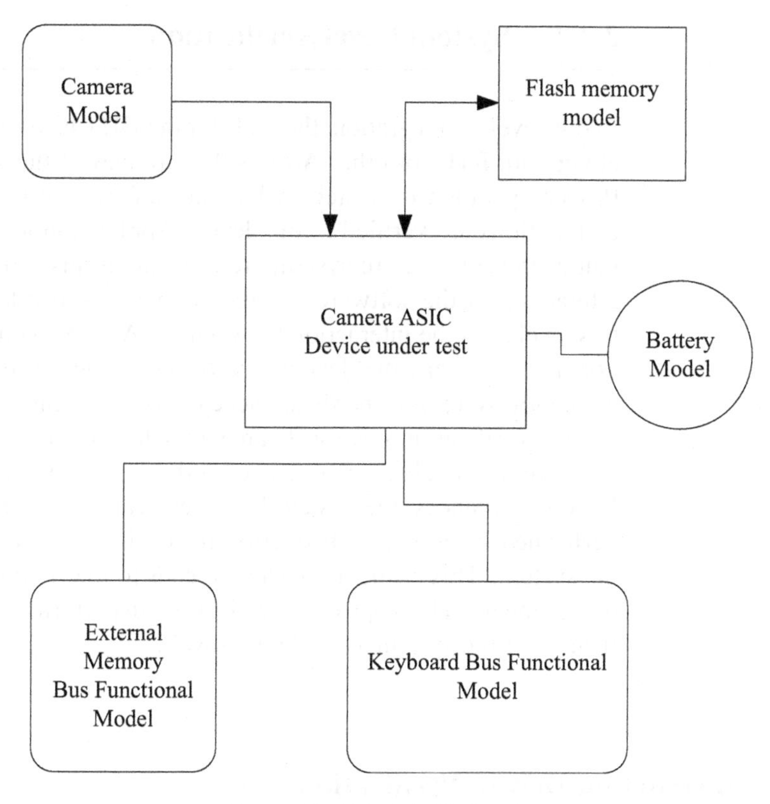

Figure 2.7. Full Chip Verification

2.3.3 Full Chip Verification

The full chip verification environment consists of the entire ASIC along with bus functional models driving the various inputs of the ASIC. The RTL is complete and there are no bus functional models or missing RTL from the device under test. The full chip environment is considerably slower than the module level environment. Full chip tests are also harder to debug given the size of the design and the slower run times. Sometimes based on the size of the design the design is regressed/verified using an RTL accelerator. Most of the tests in this level focus on end to end simulations to ensure that the entire path through the device is clean and free from bugs. In the figure 2.7 the entire chip is shown in the verification environment

2.3.4 System Level Verification

In this level of integration, the RTL for the DUT is instantiated along with RTL for other ASIC's that are part of the system. Proper operation of system behaviour and any system level assumptions are verified in this level. Application level tests that exercise the entire system are also run at this level. The interaction of the software and the hardware is also tested at this level as is the interaction between the ASICS themselves. For the most part, the design size is very large. The use of emulation systems or hardware accelerators is common at this level of verification since software simulators are hard pressed to deliver reasonable performance for this level of verification. In some instances, the system level verification is sometimes performed using some abstraction for each of the ASICS in the system. This abstraction usually leads to lower simulation requirements. This approach may be used in performance verification of the system as a whole as well.

2.4 Instruction Driven Verification

Functional verification of devices incorporating a microprocessor design has long been one of the most challenging areas in functional verification. The goal of functional verification is to achieve maximum confidence in the design. The verification challenge is compounded by the fact that there is a large verification state space for the microprocessor.

In this approach, test scenarios are coded using C-code or assembly language. The method of generating instructions varies with many approaches being adopted. The automatic method typically uses a generator that has knowledge of the instruction set architecture of the device under test. This generator may have other inputs as well (see figure 2.11 also). The generator then produces a test case or stream that is then used to validate the device under test.

The manual generation on the other hand is done by a human who manually crafts a set of test cases to address a particular scenario which may take a great deal of effort to create using the random environment. These hand-built test cases typically

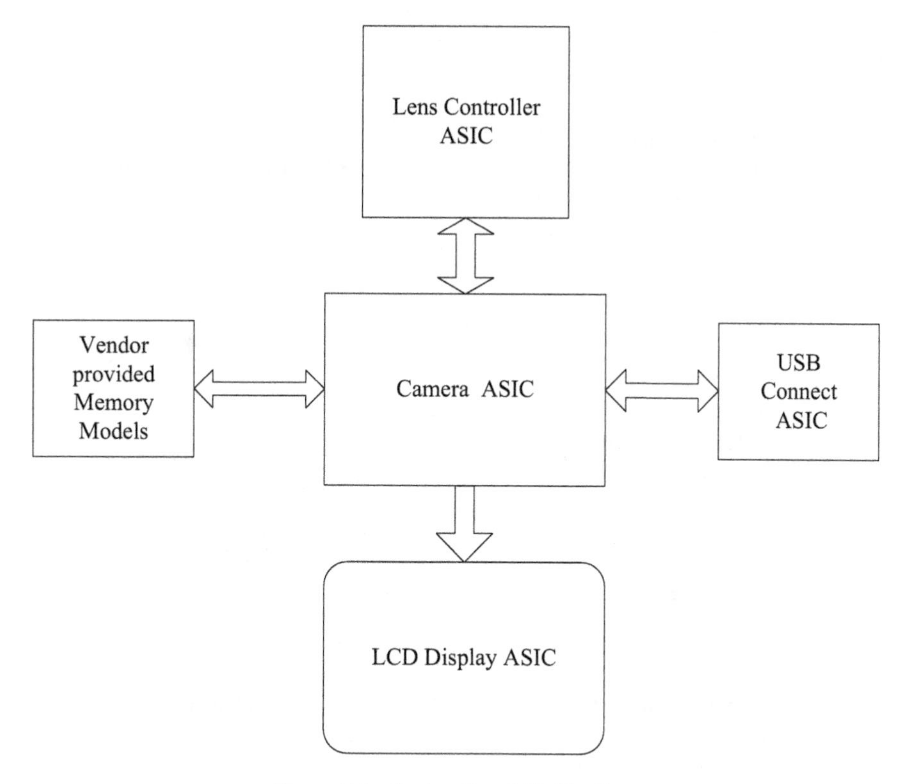

Figure 2.8. System Level Verification

augment the stream that is built using automatic generators or are used early on in the design process. This flow is depicted in the figure 2.9

Frequently, the output from this stage is also a memory image which can be loaded into the simulation world.

The simulation model picks up the image via a memory model or a bus functional model. The RTL is allowed to run through the test and execute the instructions in the test. A reference model or other method determines the correctness of the behavior of the design. During the entire verification process, many monitors are embedded in the design and in the test generators. These monitors provide various metrics that are usually used to determine the quality of the instructions generated.

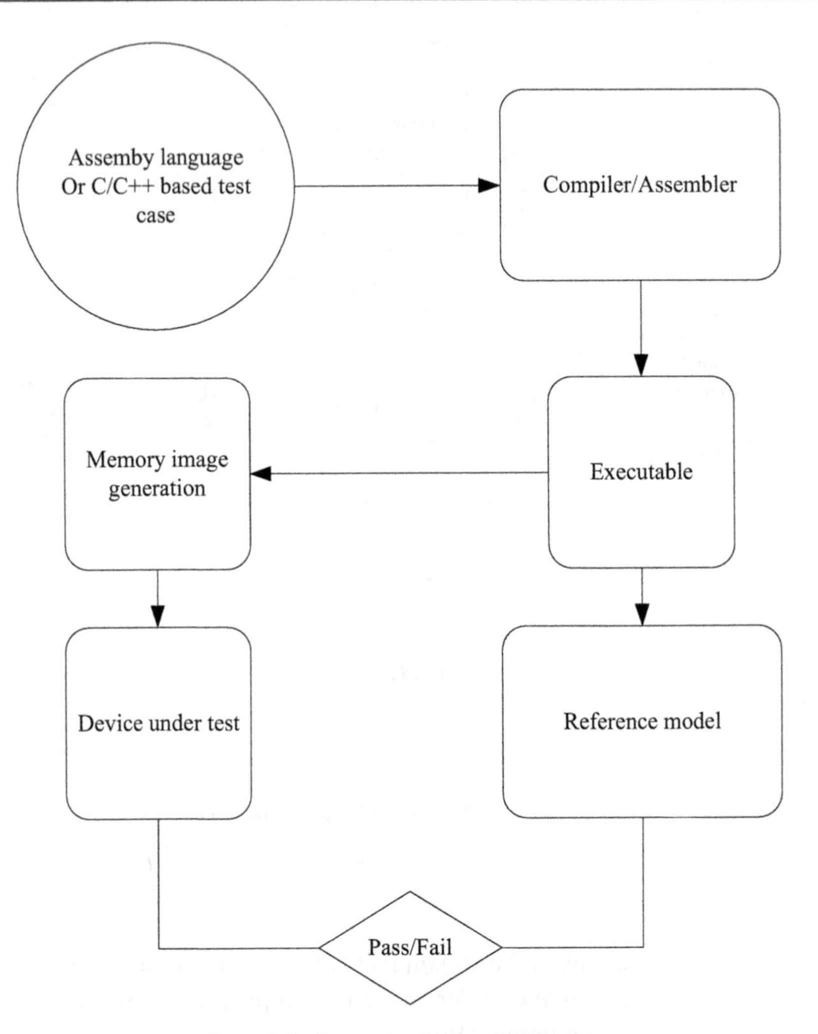

Figure 2.9. Instruction Driven Verification

This approach is very popular in microprocessor verification. There are numerous instances in literature where instruction driven verification has been deployed to create complex processors. [1],[2],[3],[4],[5] offer excellent insight into applications of this approach.

2.5 Random Testing

Random testing has gained a lot of popularity in recent times. In this approach, a test generator generates various scenarios in a test using randomness to quickly generate a variety of scenarios of interest. This approach is different from many other approaches since the approach relies on a collection of constraints to tune the random generator to produce test cases with scenarios of interest.

A typical random generator may run in parallel to the simulator or generate the data which is processed and fed into the simulation environment. A typical random generator would probably take as input, various architectural features for the device. In addition, the format for the output data along with various kinds of "knobs" is fed into the generator program. These "knobs" represent various settings to enable the random generator to produce the desired output for a given test run.The random generator then uses a variety of algorithms internally to come up with a collection of seeds that satisfy the parameters given to it.

The figure 2.11 depicts a typical random generator. (Many commercial generators would obviously have more inputs than this one! This is a conceptual description).

One of the main advantages of random testing is that it now becomes possible to uncover bugs faster than using directed tests. However, it must be realized that many a time, the random generator may not be in a position to generate specific random sequences without extensive intervention of the verification engineer. This could be due to several factors: The quality of the constraints placed on the random generator, the random generator itself and the nature of the environment and support for random test debug.

One important thing that must be noted during the use of random generators is that the output of the random generator must be continuously monitored to make sure that a large state space is indeed being generated by the random generator.

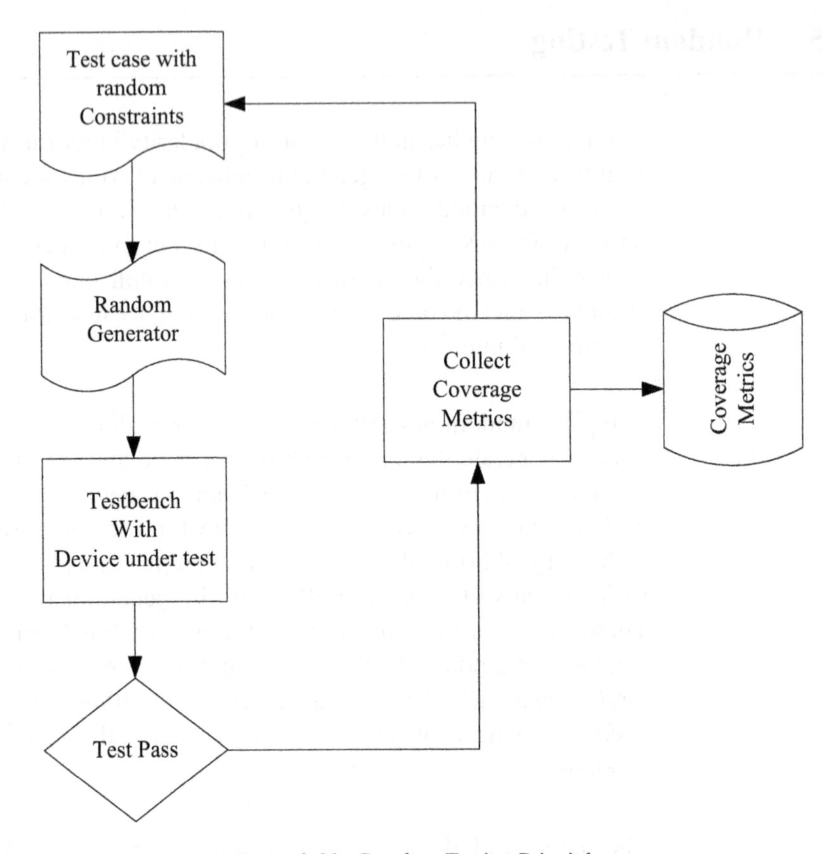

Figure 2.10. Random Testing Principles

Random testing has gained a lot of popularity with many HVL languages supporting the creation of test cases. A study comparing multiple approaches is presented in [18].

2.6 Coverage Driven Verification

With the advent of modern random generation environments, it has now become possible to use a single test to produce multiple scenarios to test the device. Coverage driven verification relies on this property of random stimuli to produce multiple scenarios automatically from a single seed. The inherent randomness of the process allows the test to uncover corner case bugs that the designer may not have considered.

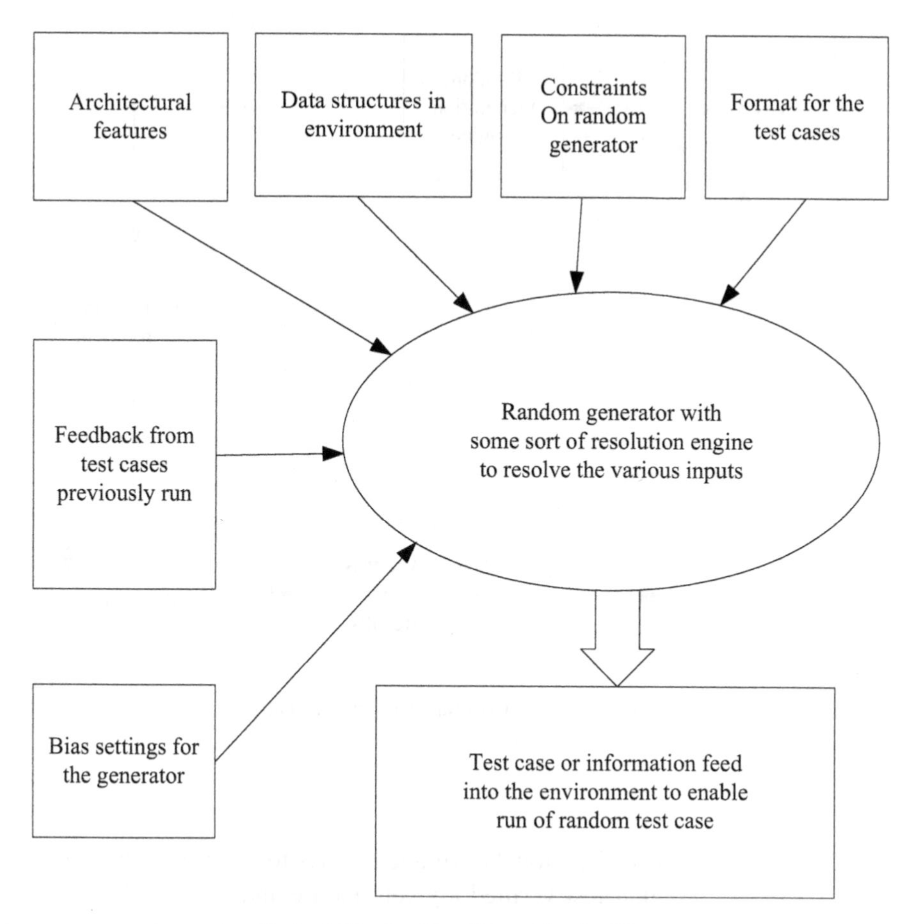

Figure 2.11. Random Generation for Tests

Coverage driven verification works by collecting coverage data. This coverage data is accumulated at various coverage points (as they are frequently called) via assertions or other mechanisms, which are exercised during the running of the test. New tests are created (either automatically or manually) to modify the constraints in an attempt to target previously uncovered features or scenarios.

A simple block diagram of a coverage driven flow is shown in the figure 2.12. In the figure, a test case with random stimuli is used to test the design. A large number of seeds is used to cover various portions of the state space in the design. Various coverage monitors are placed in the environment. The information from the coverage monitors is then used to generate more targeted stimuli for the next run.

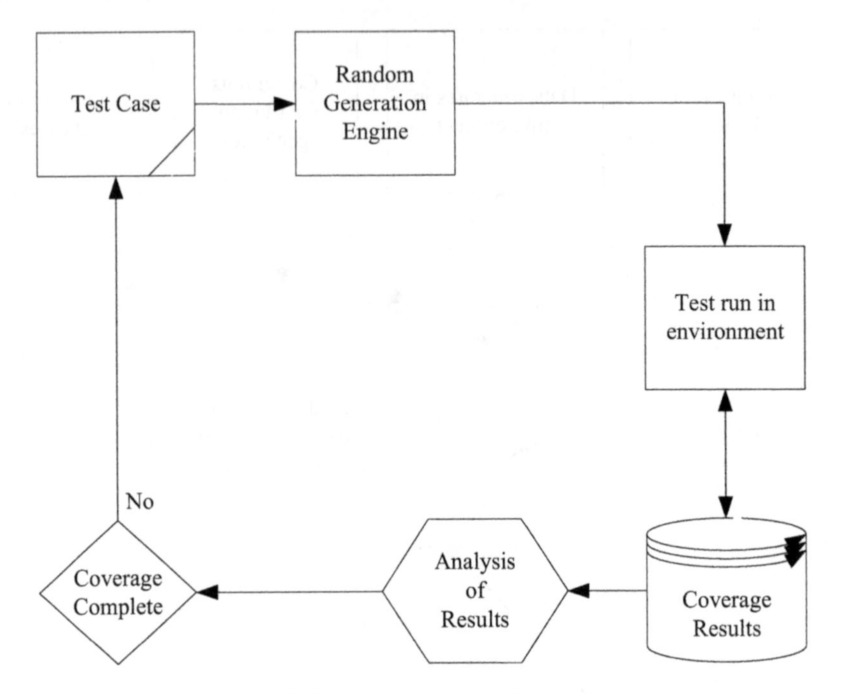

Figure 2.12. Coverage Driven Verification

Using functional coverage metrics to ascertain whether a particular test verified a given feature and feeding that information back into the process to determine the next step is termed coverage-driven verification.

Coverage driven verification has one important advantage, it helps the verification effort reach functional closure more rapidly than the directed test approach as seen in figure 2.13. A comparative graph of the coverage driven verification versus the directed test approach is shown in the figure 2.13. There is quite a bit of literature that covers this emerging and important topic in verification [1],[6],[7],[8],[9] are also good references on this topic.

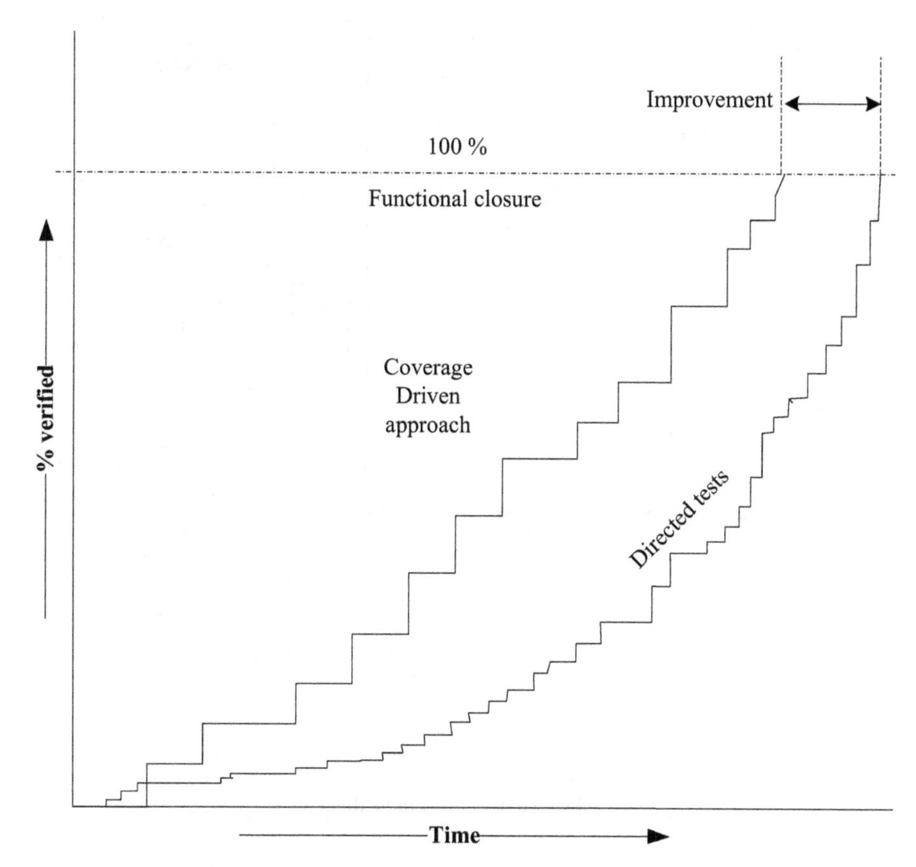

Figure 2.13. Coverage Driven Verification's Effectiveness

2.7 Transaction Based Verification

A transaction is an abstract representation of an activity in the design. Since the representation is abstract, it becomes possible that the representation will enable verification of the device with significantly lesser effort. A block diagram of a transaction based verification environment is provided in the figure 2.14.

There is a layer of abstraction built right in

Transaction level tests are tests that describe at a high level, the various transactions that should be executed. The verification environment transforms these transactions into specific behaviors that correspond to the proper abstraction level

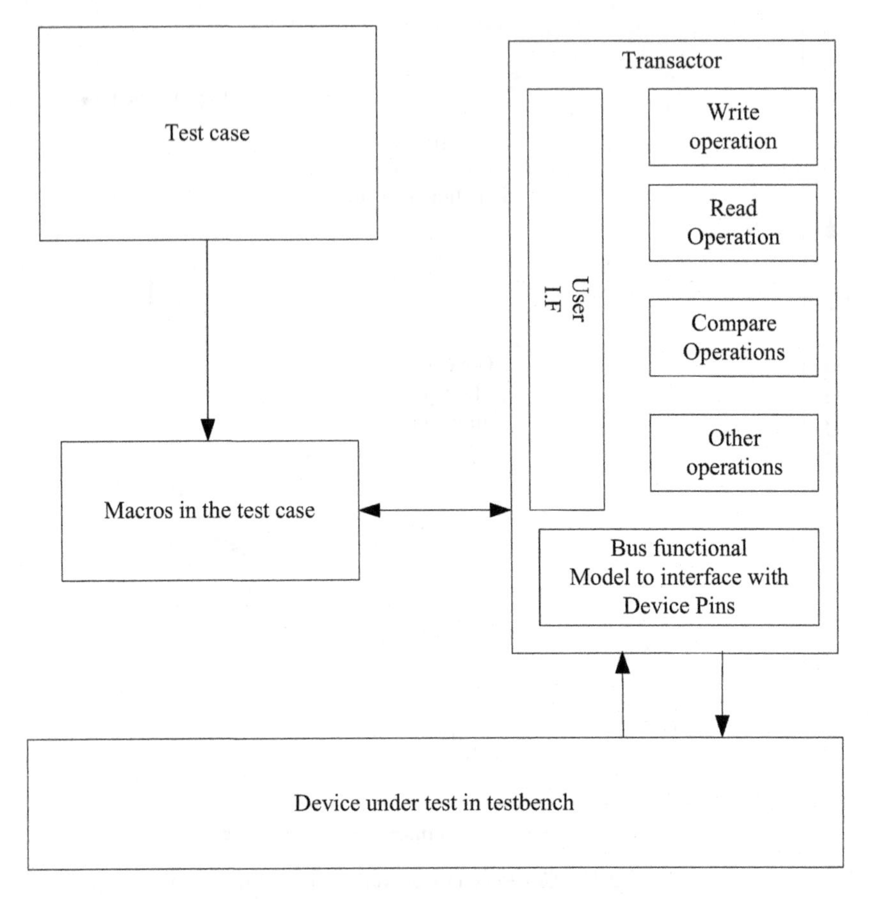

Figure 2.14. Transaction Based Verification

of the model being simulated. The transformations are usually accomplished using *transactors*, or *bus-functional models* (BFM), which perform these transformations. In the example of the bus indicated above, the actual details of the operations performed on the pins of the device are hidden from the user. The user need not worry about generating the correct protocol since the transactor will take care of such details.

A typical transactor for a microprocessor bus as indicated in the diagram may indeed consist of several components. These are also shown in the Figure 2.14. The user interface layer provides a method for the user to access the bus and perform various operations like read,write,compare data etc on the bus. There are various sections inside the transactor which implement the portions of the user interface and provide inputs to a bus functional model which then drives the signals on the bus and receives the data provided by the device under test.

There is a partitioning of the environment and the test

There is a clean division between the tests in a transaction based environment and the environment itself. The test is responsible for *describing* the transactions, while the transactors are responsible for implementing the transactions for example, toggling the right pins in the right manner etc. Hence, the environment and infrastructure can be enabled to support different levels of abstraction. Such an approach promotes modularity. Modularity is an important consideration as the complexity of of the device increases.

Coverage can be obtained easily in this approach

One interesting concept that allows transaction driven verification to become popular is that all of the benefits of coverage driven verification to be applied to the generation of transaction *descriptors*. These descriptors tell the transactor what to do and may contain a lot of information which could be randomized. Coverage may then be collected easily and analyzed.

Since transaction based verification operates by abstraction, the power of abstraction can be brought into play. Using some well defined interfaces between the descriptors and the transactors, it becomes possible to drive a high level model of the design at one point and reuse the same for driving an RTL model through a different, but compatible set of transactors. For example: Consider the cache subsystem presented in figure 2.15. The L1, L2 and L3 caches along with a PCI controller are encapsulated as the storage subsystem of a hypothetical device. In this example, the system controller and external memory are modeled by an another transactor providing interface from the PCI side to the device under test or alternatively be implemented as a bus functional model. A transactor interfaces with this subsystem and allows the user to perform various transactions to load and store the data in the cache. The state of the cache and the tags is easily determined by the sequence of operations performed on the cache by using a simple checker/monitor. The test may only consist of Load, Store, Cancel, and other operations performed by the test case. the rest is handled by the transactor. Since the user interface is simple, various operations including randomization of the address and data fields is

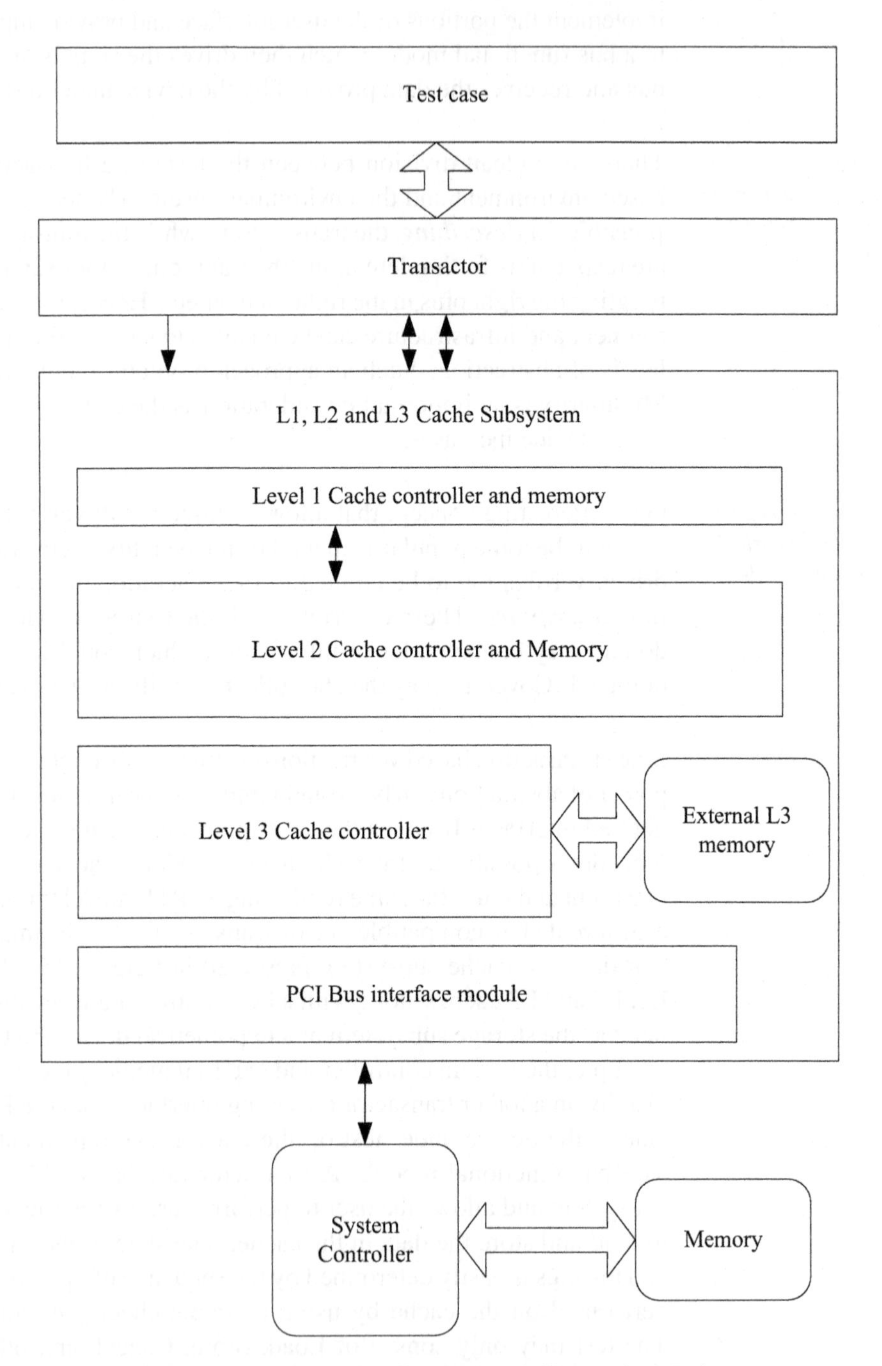

Figure 2.15. Example of Transaction Verification in a Cache

usually performed. Collection of coverage metrics is also very easy in this approach.

There are some good references on the subject of transaction based verification[2],[10]. Some commercial tools also deploy the transaction methodology as part of their offerings[14].

2.8 Golden Model Approach

The golden model approach typically uses a reference model to determine the pass or failure of a test. A test sequence is passed through the RTL and the reference model as well. The results from the golden model and the simulation are compared with one another at some predetermined points. The pass or fail of the test is determined by the fact that the results from the golden model and the design are in complete agreement. The figure 2.16 indicates a typical flow using a golden model.

Stable specification is ensured

One of the nice things about a golden model approach is that all the details of the specification are completely hashed out as the golden model is built. Every single detail is exposed to the model building process and issues are sorted out. This allows the implementation phase to be a little easier.

Specification must be clarified immediately

Since issues with the specification typically wind up holding up the golden model development, it becomes imperative that any questions in the specification must be clarified as soon as possible.

2.8.1 Advantages of a Golden Model

Automated verification is a possibility

Since there is now some sort of a checking mechanism that ensures the correctness of the device behavior, it now becomes possible for some sort of an automated checking mechanism to be used to determine the correctness of the device under test. This leads to the concept of running random regressions with random seeds and only looking at the tests that failed!

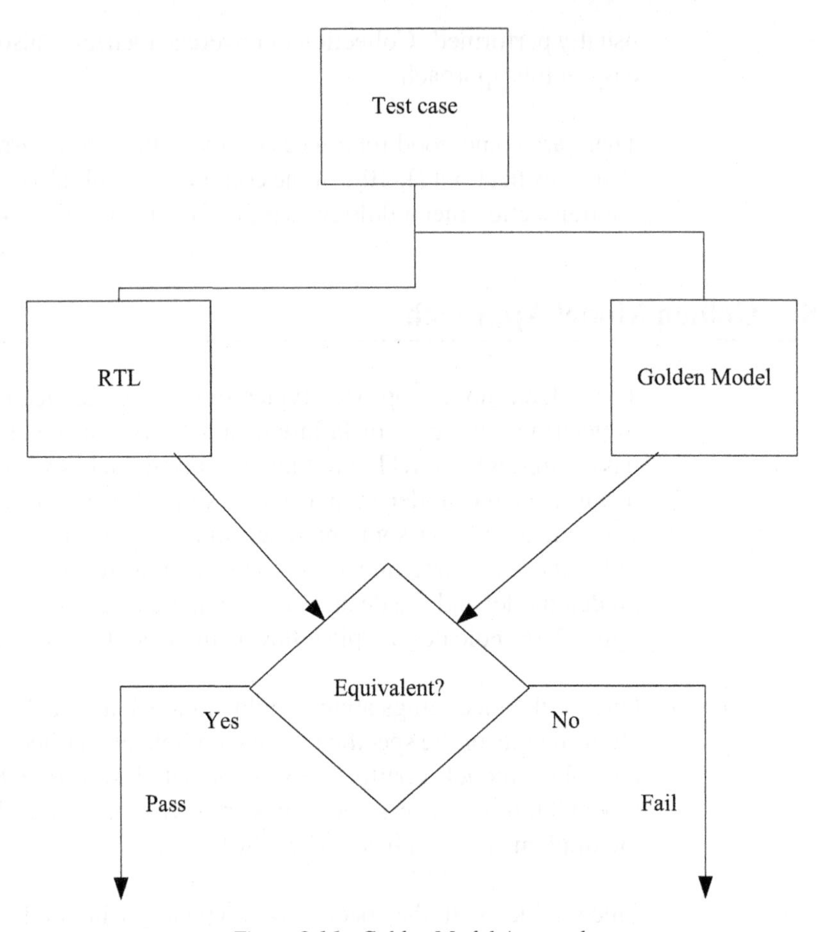

Figure 2.16. Golden Model Approach

Defects are identified immediately

Another nice thing about golden models is that it allows the user to identify defects immediately. The failure detection happens at the point of failure since there is a mismatch.

2.8.2 Disadvantages of using a Golden Model

Reference is to be modeled

The biggest concern with a golden model approach is that the model needs to be built. Many a time, this time consuming process in some organizations may take many months to complete. The model development activity can sometimes be as large as the RTL development activity.

Initial effort is higher

Creation of the golden model requires some work. Depending on the initial size and complexity, the golden model could take a lot of time as mentioned earlier. In addition to the golden model development, the verification environment has to be created with the right "hooks" to be able to take advantage of the golden model.

In some circumstances, the entire device may not be available as a golden model. Only a portion of the device may be available. For the other portions, a different approach will be needed.

In addition, golden models come in all shapes, sizes and packages. Some of the models are transaction accurate, some others may be cycle accurate, and others accurate for a specific set of inputs and outputs. It all depends on what is available in the time frame allotted to the design of the device and what is available from the golden model.

Bugs!

Another major issue is that the golden model too has the possibility of bugs. A failure that is not flagged does not imply that no error exists. If the specification changes significantly, then a major rework of the golden model is in order. The golden model also needs to go through regressions using test cases that are created to verify the correctness of the golden model before it is completely usable. The presence of a golden model also implies that some verification environment needs to be built around the golden model to make sure that the error evaluation and reporting happens at the appropriate time.

There has been considerable debate on the pros and cons of a reference golden model. However, as discussed earlier, there is a significant advantage in using a golden model for parts of the design if it were available. It does speed up verification significantly.

2.9 Pre-Post Processing Approach

Pre-Post processing approaches have found application where there is some sort of a complicated program or algorithm that makes it challenging to set up a self checking simulation.

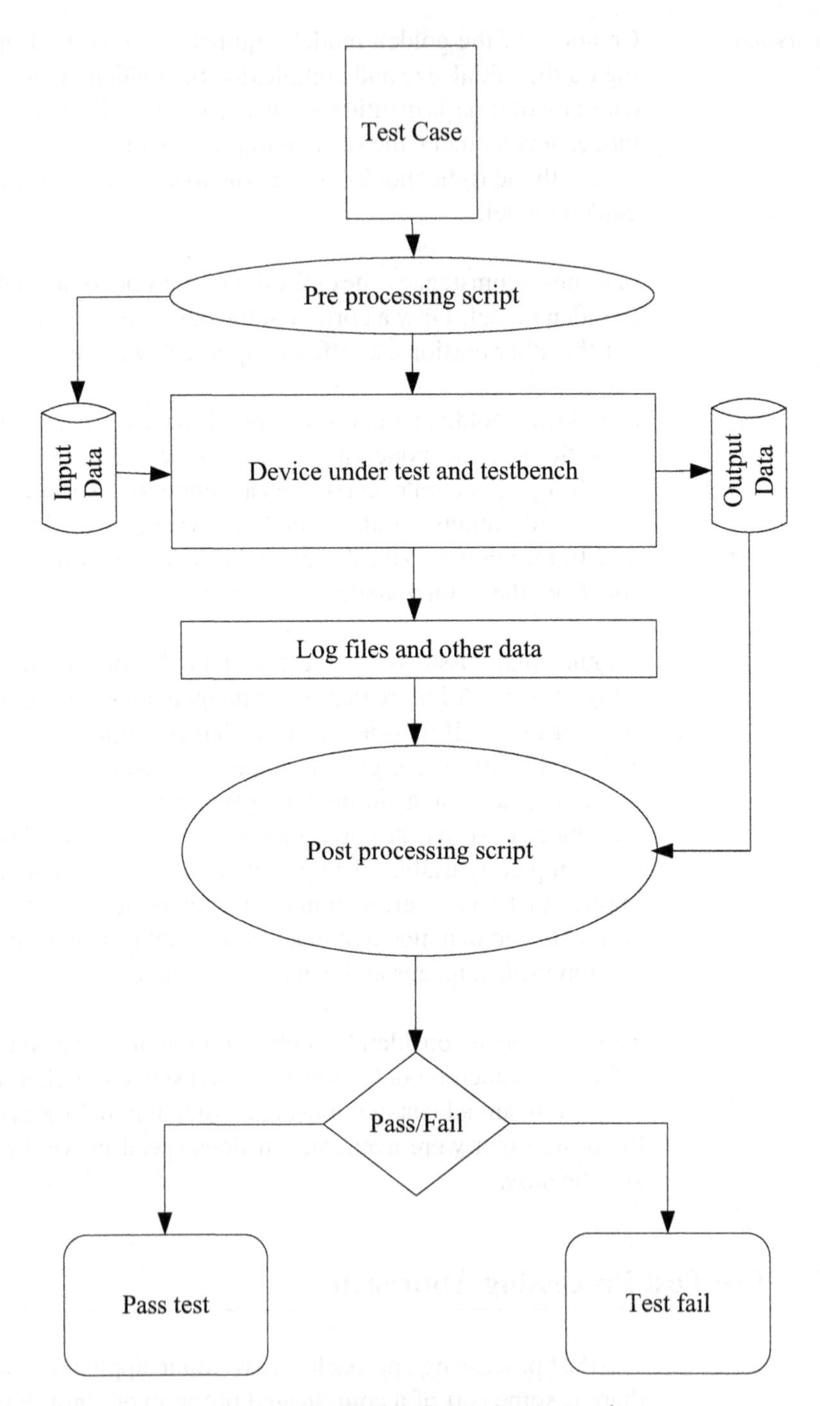

Figure 2.17. Pre-Post Processing Approach

In this approach, a script or a program generates input data based on some input constraints and saves the data in a file. This data could be data which is marked using timestamps or other information as well. The testbench reads the data from the file and then feeds it to the device under test. The test bench may also choose to perform some other operations on the design when the data from the file is being read in. The data from the device under test which is collected by the test bench is then saved in another file. A script is then run on the collected data and a pass or fail is determined by the post processing program.

This might be a good approach in some cases

This approach works well for data driven types of testing. The other advantage of this approach is that there is the possibility of the use of a standardized third party program to help with the data analysis. (Standards compliance etc.) In many circumstances, the post processing script is an expensive program in terms of cost/license. The pre-post processing approach allows the optimal use of the licenses of such programs since only a small number of licenses are needed for the verification environment.

Real time failure detection is a challenge in this approach

In some organizations, this approach is called as a "Store-replay" approach.The approach seems to work well in situations where there is Black Box testing approach being adopted. (Standards compliance or similar situations) One of the main disadvantages of this approach is that feedback in the test bench is not real-time. IE: The test or the test bench cannot determine and terminate the test at the point of failure. This is due to the inherent limitation that the failure metrics are outside the test bench. Hence, if a test is a long one, and the failure happens early on in the test, the entire run is wasted. However, other monitors and checkers may help alleviate this issue.

The author has successfully used this approach in a couple of organizations when the device under test had some very complex frames. The post processing program was a third party tool that did lend itself to co-simulation. However, we chose this approach since the licensing requirements for the analysis tool was kept at a minimum.

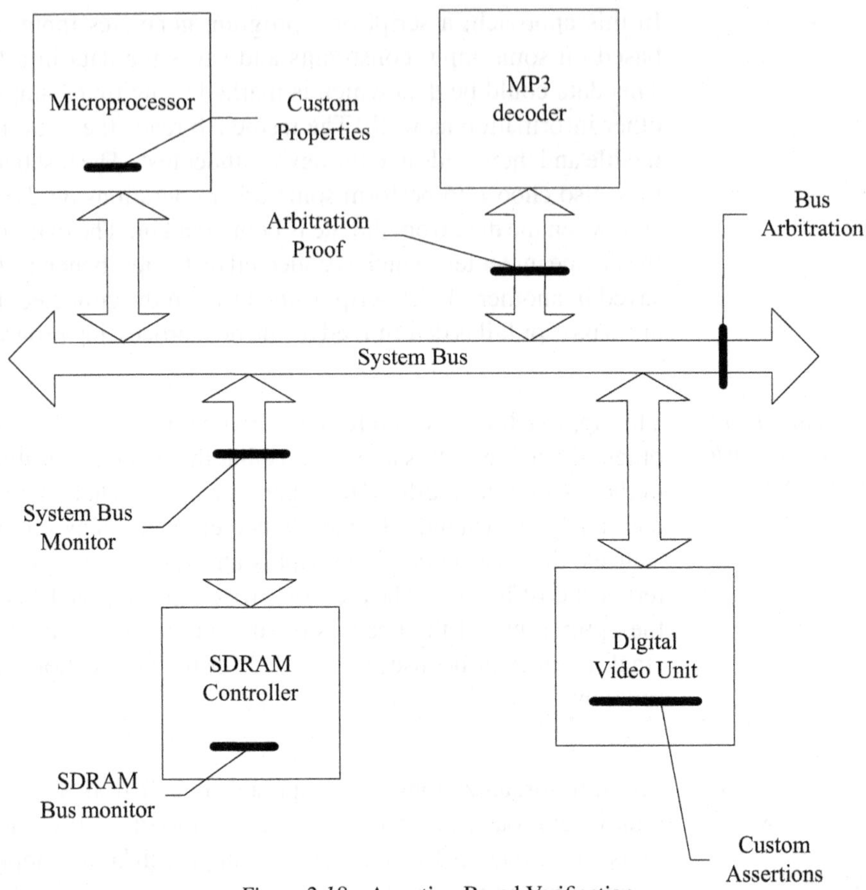

Figure 2.18. Assertion Based Verification

2.10 Assertion Based Verification

One of the newer methodologies that has found adoption recently is assertion based verification. This methodology uses assertions embedded into various portions of the design to help speed up the verification process.

An assertion is concise description of a complex or expected behavior. Assertions can be specified using specification languages like PSL or using OVL modules,SVA or other approaches. Many vendors have their proprietary formats as well.

Assertion based verification makes use of a library of assertions that are embedded in the design. This library has a few

components that make the library easy to use. Simulation is then carried out with the assertions which trigger based on their inputs. Hence, considerable value is rapidly provided with little or low effort. Coverage metrics using assertions can also be realized in this approach. This methodology has an advantage in that it works all the way from block level through to chip and system level.
There are many ways to use assertion based verification in HDL. Many vendors support different flows in assertion based verification. Some of the common approaches are:

- Development of assertion IP in HDL and then plug them into the HDL design. For example: (CheckerWave library),0-In
- Develop language to write assertions, provide interfacing with design. For example: PSL(sugar)[11], TNI-Valiosys.
- Assertion modules are written in in existing HDL and then instantiated into the design. For example: OVL (Verplex BlackTie UDC)[12].
- Extend existing high level verification language for introducing lower level assertions. For example: OVA(Synopsys).
- Extend language to provide assertion capability in itself. For example: System Verilog(SVA).

Since the assertions are placed in the RTL, it becomes possible to deploy the same assertions for formal verification. Overall benefits of using assertion monitors are outlined below:

- Provides higher abstraction for more visibility in dynamic simulations.
- Simplifies the diagnosis and detection of bugs (if placed closer to 'potential' origin of the bug)
- Increased coverage of the design.

Assertions can be used in many places. A brief diagram which indicates where the assertions are possible is shown in the figure 2.18. The list presented below is not an exhaustive list

- Protocol checker assertions help ensure that the protocol of the bus is not violated.
- Arbitration assertions make sure that the bus is arbitrated for properly and any improper decisions are flagged.

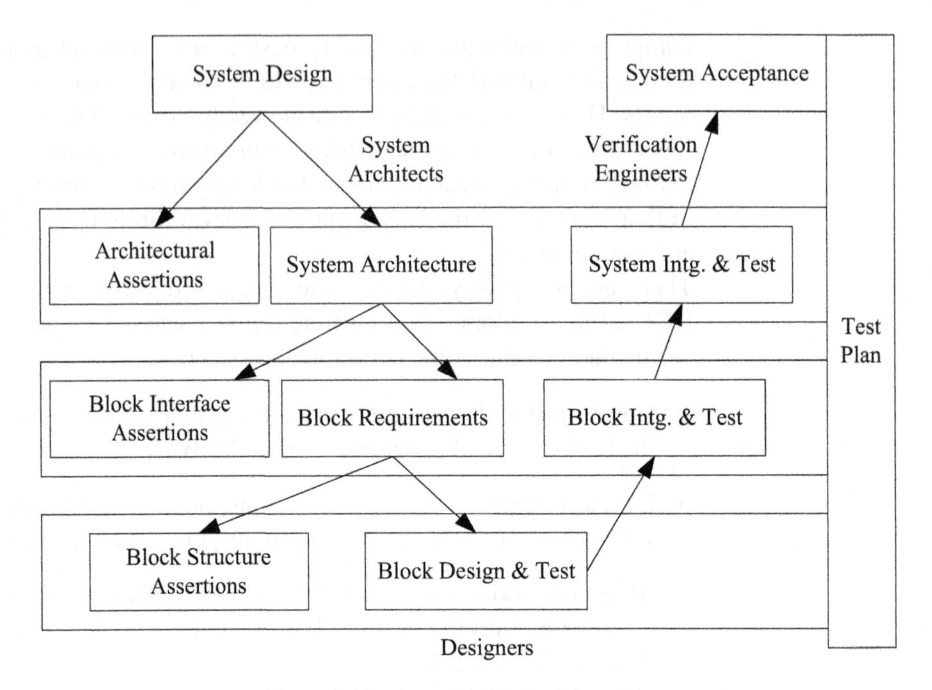

Figure 2.19. Assertion Methodology Flow

- Custom assertions deep in the design ensure proper operation of the module.
- Specific properties of the designs to be proved by assertions
- Some assertions also act as gates or constraints helping invalid inputs from being presented to the design.

2.10.1 Assertions - Who writes them and when?

Architects

Architects can use assertions to describe high level relationships. These relationships could describe events that ensure system level behavior or ensure system-level consistency. The architects can use the assertions to ensure that the assumptions that they made when they designed the system are not violated when the system is designed.

Design Engineers

Design Engineers on the other hand are left with the task of ensuring that the design is implemented. The assertions can be used to capture design understanding as well as interface assumptions that they make when they create the design. Any specific corner cases can also be captured by the designer. Assertions must capture designer's understanding for any specific implementation intentions or restrictions.

Verification Engineers

Verification Engineers are concerned on the correctness of the design. The assertions can be used in test benches or as "early warning" indicators. Assertions can also serve double duty for coverage monitors or as checkers. In addition, assertions can serve as properties to be proved or as constraints in formal verification.

The various levels at which assertions can be used is presented in the figure 2.19.

2.10.2 Types of Assertions

Assertions have both a combinatorial flavor as well as a temporal flavor [13]. This is what makes them extremely attractive to deployment.

Static assertions

Static assertions must always hold true. They don't have any time related properties and must be true for all time. For example: Signals A and B cannot be different at the same time as shown in 2.20. These assertions usually have some sort of combinatorial logic associated with them. They are evaluated all the time.

Temporal assertions

Temporal assertions on the other hand are valid at specific time instances in the design.They are triggered by the occurrence of some specific event in the design. Temporal assertions act over a period of time. For example: After signal A is true, then signal B must be asserted 1-3 clocks later. This is indicated in figure 2.21.

2.10.3 Advantages of an Assertion Based Methodology

Exhaustive block/module level testing

Using assertions in the methodology enables formal verification. It becomes possible to formally verify the block or module

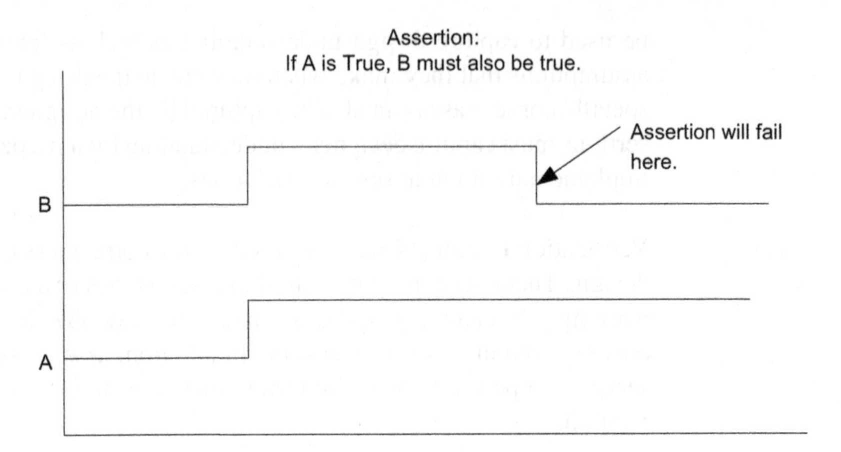

Figure 2.20. Static Assertions Example

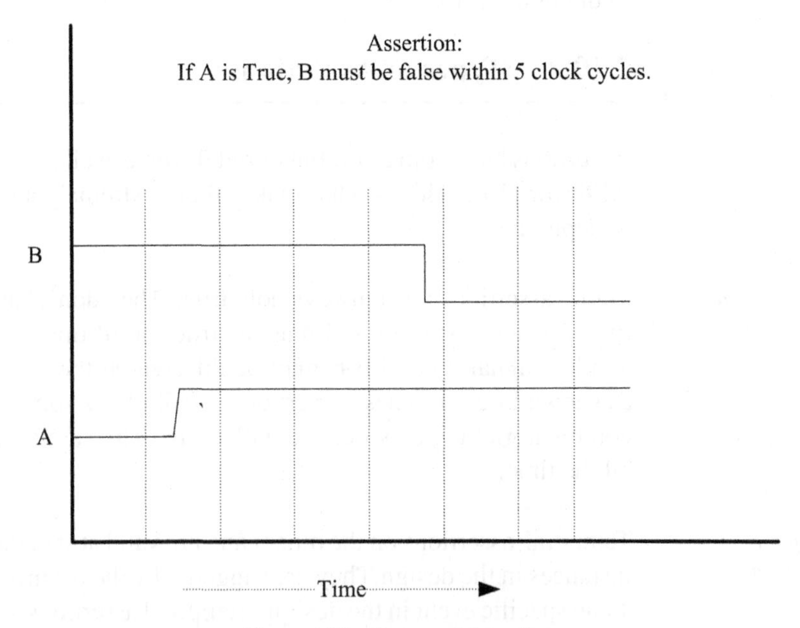

Figure 2.21. Temporal Assertions

level using available approaches. A formal analysis tool can use the assertions and determine statically if the design does indeed meet the constraints and properties specified by the assertions.

Interface or Protocol compliance check (integration)

Block/system level integration of designs that use assertions is greatly simplified since the assertions typically help ensure that the interface is indeed operating correctly and valid inputs are available to the module that is being integrated.

Reduced debug time

Assertions when properly placed can give a good indication of where the problem is. Properly coded assertions with appropriate messaging can be used by the test environment to terminate the test as soon as a failure occurs by detecting messages in a log file.

Increased confidence in the design

Another important aspect of verification using assertions is that it helps to ensure that the intent of the design is met at multiple levels. At the architectural level, the assertions ensure that the high level goals are met. At the design level, assertions help ensure that the intent of the designer is met. The verification assertions help to identify problems quickly.

Always there

Once the assertions are inserted into the design, they are present in the design for the lifetime of the design. The assertions will be present even if the design is reused in another project. Hence, it becomes possible to communicate the designer intent to other project members who may work on the design later.

Self documentation

One of the biggest advantages of instrumenting a design with assertions is that the code is now effectively self-documented. This is particularly important as the code evolves through the design cycle. The code is now portable to some other project. The code is explicit and contains all the assumptions and designer's intentions. It becomes easier to read and understand well documented code.

Simulation with assertions

Assertion based verification brings together design and verification to improve both the code and the verification process. By adding assertions during the coding phase, designers get value throughout the process of design, integration, and full device simulation.

One major design team reported recently that assertions used throughout the design cycle could be credited for finding the majority of bugs not found by other techniques.

2.10.4 Challenges with an Assertion Based Methodology

Not all designers want to use assertions

Many designers may feel that the assertion based methodology may not buy much in terms of advantages. Many of them also feel an additional workload that is imposed on them in the face of already tight schedules.

Designers who experience assertions believe it is a waste of time

Many a time, the designer is left to debug an assertion that fails when a simulation is run. Often the designer may discover that the failure is due to an incorrect coding of the assertions. Many designers also feel that the maintenance of assertions is also an issue. One concern is that defective assertions will repeatedly fail in the design. The assertions for the feature being verified could be wrong, or incorrectly written. In either case, significant verification time is wasted.

How many assertions are sufficient in a design ?

Another challenge is the question of the number of assertions in the design. It is crucial to ensure that every assertion 'pays for itself' in terms of the number of bugs found as a result of the assertion being present in the design. Adding excessive assertions quickly becomes a counterproductive effort. Hence a balance must be struck between the number of assertions and the quality of the assertions in the design. A method of deriving the important assertions is presented in the chapter *Putting it all together*.

Danger of simulation vs. silicon mismatch if 'checker' code not between 'synthesis translate on/off directives

Due care must be exercised when using assertions since the checks that may contain synthesizable code as part of the assertion itself. This is particularly true of assertions developed in Verilog, VHDL or the assertion library OVL.Potential for a problem exists if the code for the assertion is accidentally rendered on silicon if some additional "glue" code were to be developed to help insert the assertion in the design.

Need mature tools and methodology to proliferate its use

Assertions are still a new methodology from recent times. A number of approaches are being developed to use the full power of assertions. A variety of homegrown tools are required to be able to mine data from assertions.

Assertion based methodologies have gained a lot of importance in recent times. There is quite a bit of literature now available on this topic as evidenced in [11], [13], [14], [15], [16], [17].

2.11 Formal Verification

Formal verification is an emerging technology in recent days that has helps to verify designs rapidly. Formal proving engines have existed for many years; However, their application to solve functional verification challenges is recent.

The focus of the formal verification methodology is primarily to ensure that the intent of the design is met using assertions that are embedded in the design and test environments. It focuses on proving that the designs architectural and structural intent are completely proved using a mathematical process rather than a simulation based approach where test cases are driven on the design. Formal verification is a complete mathematical proof.

Formal verification in the author's classification is an extension of the Assertion Based Methodology presented in the previous section. Formal verification attempts to prove that the assertions embedded in the design can never be violated or alternatively finds a a stimulus sequence that violates the assertion. This violation is termed as a "counter-example". If neither proof or counter-example can be found then assertion is termed to be "indeterminate" in the formal approach.

One of the main advantages of a formal tool is that it does not need an extensive amount of verification knowledge and provides results rapidly. The tools do not need extensive test benches and lend themselves well to (block)/module level verification. The designer can add assertions to the design to verify design intent and formally prove that his assertions are true. The tool proves either that the assertions are correct or provide a counter example stating where the assertions fail.

Formal verification is typically classified into two areas.

- Static Checking
- Dynamic Checking

Static formal verification

Static checking uses mathematical techniques used to prove an assertion or property of the design. It is an exhaustive approach for complete logic analysis and coverage. A formal verification tool reads in the assertions and the design and attempts to completely prove the assertions. There is however a capacity problem with the current generation of tools which must be addressed. This capacity problem is currently posing some challenges to deployment of these tools.

Dynamic formal verification

Dynamic formal verification on the other hand uses static capabilities as well as simulation. This approach attempts to supplement simulation results based on bounded model checking (BMC) algorithms that amplify existing simulation tests for generating counter-examples.

Simulation with assertions exposes bugs that are stimulated by existing diagnostic tests as the assertions are run in simulation in addition to running in the formal tool. Dynamic formal verification with partial constraints quickly finds bugs by leveraging simulation results.

Deep Dynamic Formal simulation

Deep Dynamic formal simulation is a third method that has come about recently. It is similar to dynamic formal verification and targets finding "counter-examples". However, it is based on ultra-fast BMC or similar algorithms, with much larger proof radius and is said to be capable of verifying very large designs.

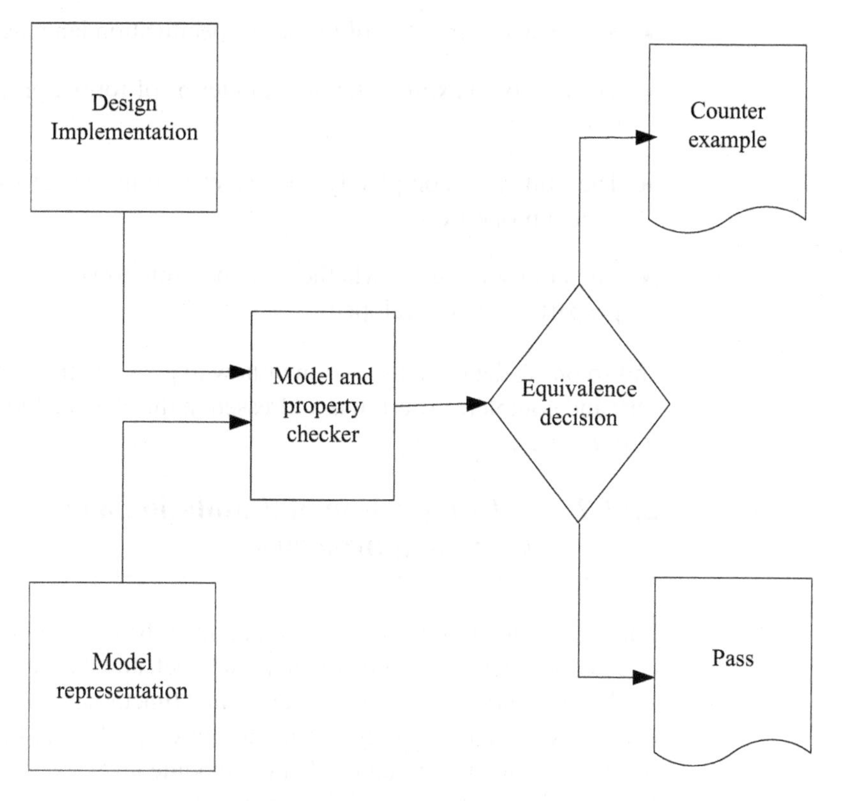

Figure 2.22. Formal Verification Model Checking Approach

2.11.1 Model Checking

Inthe model checking approach, The design intent is expressed in terms of formal properties. A formal check verifies whether the implementation satisfies these properties. If this indeed true, the checker reports success, otherwise it produces a counter-example. Such an approach is shown in the figure 2.22

There are several challenges to using property checking. However, when compared to the simulation world, the behavior is exhaustively verified in a formal verification approach. There is a mathematical representation of the specification created in a formal approach which is then verified against the constraints and properties. However, promising as it may appear, the formal approach today quickly runs into capacity and other challenges as defined under the following broad categories.

- Syntax and semantics of property specification languages.
- The ability to express the intent in terms of formal properties.
- The ability to completely specify everything in terms of formal properties.
- The ability to verify whether it is possible to verify every property that is developed.

and so on. A detailed discussion of how to generate the necessary and complete criterion is addressed in the chapter *Putting it all together*.

2.11.2 A Comparison of Simulation and Formal approaches

Formal verification is a bit of a paradigm shift for verification engineers. There is no concept of a test bench and the standard well known metrics like code coverage and functional coverage do not exist. The capacity of the current crop of tools limits their deployment to being used at the module level. Nevertheless, the ability of being able to verify a state space completely very quickly offers a significant advantage over conventional approaches and must be considered appropriately as a part of the verification arsenal.

Simulation has traditionally been the workhorse of the verification effort. The design under test is integrated into an environment which then passes a variety of vectors through the design. Over a a period of time, there have been various metrics that have evolved to help gauge the state of the verification effort.

Nowadays, designs instantiate assertions as part of the design and use the same in the formal verification approach. There are many languages today that can be used to describe the assertions based on the readers choice and environment used to design and verify the device. In recent past, there have been several developments in the field of formal and advanced hybrid functional verification. This is well evidenced by the current state of the art. Formal tool research of date focuses mostly on a couple of main thrust areas: The capacity of the formal engine by means of various algorithms and the concept of "intent" where the level of abstraction is raised and the intent of the designer is proven.

Unfortunately, both these approaches described above are plagued by their inability to handle a large design. Varying results have been reported at various companies which have used a variety of techniques to enhance the size of the design that can be handled by the formal tool. It is observed that depending on the coding style used and the number of gates that are inferred, the tool rapidly approaches a capacity limit. When this occurs, the tool typically results in an "inconclusive" report detailing the progress and results the tool was able to achieve.

However, "inconclusive" reports have very little meaning to the user of the tool. Such results are effectively imply that there is the need for extensive work on the part of the user to either prove the assertions somehow or abandon the effort in favor of a different approach. There is no counter example in this case.

Currently, most research is in enhancing the capacity of the tool and its usability, Many flavors of formal verification tools are available in the market nowadays with different capabilities and approaches.Many companies now use formal verification at some level to help verify their designs.

On the other hand, simulation based approaches which are well known today suffer from other issues of their own. Extensive time is usually invested in simulation, test development and debug. The environment does allow engineers to think in a serial fashion and allows progress to be made albeit slowly. Simulation provides coverage and other metrics which act as confidence building measures before taping out a design. In spite of these advantages, it is hard to prove complete verification of anything. One relies on metrics to determine closure.

The choice of an approach is therefore determined by a variety of factors including the size of the design amongst other factors described elsewhere in this book.

2.12 Emulation and Acceleration

Given that design complexity is rapidly increasing, there is now the need to be able to simulate RTL rapidly and ensure that the hardware-software interface is indeed working as designed in contemporary designs.

With the rising complexity of contemporary devices, it is now a challenge to use a software based simulator to provide the throughput required to run some large applications on the RTL design.

This section provides a brief description of the various emulation and acceleration related technologies. While emulation/acceleration may not qualify as an approach to verification, the description below is provided in the hope that the verification engineer may use this to speed up simulations by adapting to the flows presented by these technologies.Emulation and Acceleration are two distinct technologies as described below.

Emulation

Emulation maps the design into a device that speeds up the RTL simulation by running the design on some specialized hardware. A software-hardware interface is built onto this system allowing the user some flexibility as well.

Acceleration

Acceleration On the other hand compiles the design into a executable which is run on a specially designed CPU. This approach offers greater performance over conventional software simulations run on general purpose microprocessors. Similar to the previous approach, a software-hardware interface is built onto this system allowing the user some flexibility as well.
The basic flow for both approaches is shown in the figure 2.23. The RTL and the testbench for the device are both compiled into a format that is acceptable to the emulation device. This compilation happens via software that accompanies the emulator. The compiled netlist is then downloaded into the emulator. Tests are then run on the emulator which acts as a very fast simulator.

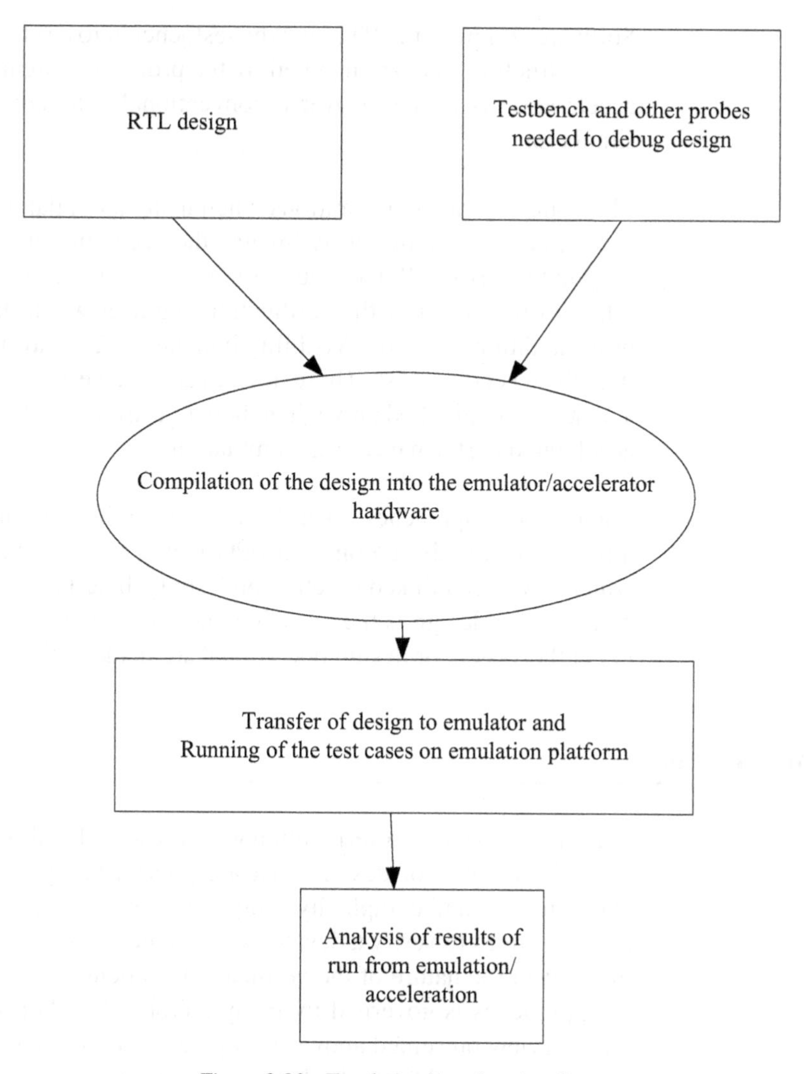

Figure 2.23. Emulation/Acceleration flow

FPGA based emulators

Hardware accelerators of this kind typically are made up of a collection of FPGA devices that are built into some sort of a matrix. The compilation software maps the RTL design into many FPGA's and partitions the design so that the design will fit on the FPGA's in the system. Speeds of a few megahertz are reported on such systems.

Processor based accelerators

Processor based devices on the other hand have one or more processors which are hooked up into a special configuration.

Software compiles the RTL and the testbench into a set of special instructions that are handled by the processor. Significant acceleration is achieved over a conventional software based simulation.

There are several considerations when using a emulator or an accelerator. These primarily involve the accessing of signals deep in the design. Some emulation devices may need recompiling of the design if the verification engineer would like to probe additional signals. Visibility into the design is available, though at a certain cost. These devices are also very expensive and hence are justified only when there is a strong need to have accelerated performance from simulation.

Some other approaches taken by companies include placing a portion of the design onto an FPGA or a group of FPGA's which have been linked together on a single board or multiple boards. The design is then verified and any changes required to get the system or design operational are made to the design.

Conclusions

Various approaches to verification were presented in this chapter. The number of designs that are present today are very varied in type and complexity. Sometimes, more than one approach is used to verify a device as was indicated in the example. Given the nature of the verification challenge, the choice of approaches is governed by many factors. It is hoped that the overview presented above allows the reader to identify the approach required to verify the design appropriately.

References and Additional reading

[1] Benjamin, Mike, Geist, Daniel, Hartman, Alan, Mas, Gerard, Smeets, Ralph, and Wolfsthal, Yaron (1999). A study in coverage-driven test generation. In *DAC '99: Proceedings of the 36th ACM/IEEE conference on Design automation*, pages 970–975, New York, NY, USA. ACM Press.

[2] Brahme, Dhananjay S., Cox, Steven, Gallo, Jim, Glasser, Mark, Grundmann, William, Ip, C. Norris, Paulsen, William, Pierce, John L., Rose, John, Shea, Dean, and Whiting, Karl (2000). *The transaction based methodology*. Cadence Berkeley Labs, Technical Report # CDNL-TR-2000-0825, San jose, California, United States.

[3] Bentley, Bob (2001). Validating the Intel Pentium 4 Microprocessor. In *DAC '01: Proceedings of the 38th conference on Design automation*, pages 244–248, New York, NY, USA. ACM Press.

[4] Lee, Richard and Tsien, Benjamin (2001). Pre-silicon verification of the Alpha 21364 microprocessor error handling system. In *DAC '01: Proceedings of the 38th conference on Design automation*, pages 822–827, New York, NY, USA. ACM Press.

[5] Malley, Charles H. and Dieudonn, Max (1995). Logic verification methodology for PowerPC microprocessors. In *DAC '95: Proceedings of the 32nd ACM/IEEE conference on Design automation*, pages 234–240, New York, NY, USA. ACM Press.

[6] Grinwald, Raanan, Harel, Eran, Orgad, Michael, Ur, Shmuel, and Ziv, Avi (1998). User defined coverage - a tool supported methodology for design verification. In *DAC '98: Proceedings of the 35th annual conference on Design automation*, pages 158–163, New York, NY, USA. ACM Press.

[7] Ho, Richard C. and Horowitz, Mark A. (1996). Validation coverage analysis for complex digital designs. In *ICCAD '96: Proceedings of the 1996 IEEE/ACM international conference on Computer-aided design*, pages 146–151, Washington, DC, USA. IEEE Computer Society.

[8] Lachish, Oded, Marcus, Eitan, Ur, Shmuel, and Ziv, Avi (2002). Hole analysis for functional coverage data. In *DAC '02: Proceedings of the 39th conference on Design automation*, pages 807–812, New York, NY, USA. ACM Press.

[9] Piziali, Andrew (c2004). *Functional verification coverage measurement and analysis*. Kluwer Academic Publishers, Boston.

[10] Kudlugi, Murali, Hassoun, Soha, Selvidge, Charles, and Pryor, Duaine (2001). *A transaction-based unified simulation/emulation architecture*

for functional verification. In *DAC '01: Proceedings of the 38th conference on Design automation*, pages 623–628, New York, NY, USA. ACM Press.

[11] Cohen, Ben, Venkataramanan, Srinivasan, Cohen, Ajeetha Kumari(2004). *Using PSL/Sugar for formal and dynamic verification : guide to property specification language for assertion-based verification*. vhdlcohen publishing.

[12] Open Verification Library, www.accelera.org. *Open Verification Library*.

[13] Foster, Harry, Krolnik, Adam, and Lacey, David (c2003). *Assertion-based design*. Kluwer Academic, Boston, MA.

[14] Yeung, Ping (2002). *The Four Pillars of Assertion-Based Verification*. www.mentor.com/consulting, San jose, California, United States.

[15] Yeung, Ping (2003). *The Role of Assertions in Verification Methodologies*. www.cadence.com/whitepapers, San jose, California, United States.

[16] Synopsys Inc(2005). Hybrid formal verification.

[17] Wang, Dong, Jiang, Pei-Hsin, Kukula, James, Zhu, Yunshan, Ma, Tony, and Damiano, Robert (2001). Formal property verification by abstraction refinement with formal simulation and hybrid engines. In *DAC '01: Proceedings of the 38th conference on Design automation*, pages 35–40, New York, NY, USA. ACM Press.

[18] Bartley, Mike G., Galpin, Darren, and Blackmore, Tim (2002). A comparison of three verification techniques: directed testing, pseudo-random testing and property checking. In *DAC '02: Proceedings of the 39th conference on Design automation*, pages 819–823, New York, NY, USA whitepaper8001. ACM Press.

[19] (2001). *Testbuilder/SystemC verification library*. www.testbuilder.net, San jose, California, United States.

[20] Dr Andreas Dickmann Lessons learnt: system verification methodology and Specman e. In *ClubV '03: Proceedings of the Club verification Europe*

Chapter 3

VERIFICATION WORKFLOW PROCESSES

Various Workflows Practiced in Verification

In this chapter, a brief overview of the various processes that occur during the verification cycle are presented. Some organizations follow these processes with some degree of formalism; Other organizations enforce these processes in an informal manner. These processes vary in terms of detail or the exact stages from organization to organization. The terminology also sometimes tends to vary. The principles behind these processes however,are common to all organizations.

3.1 An Overview of the Entire Verification Process

Today's complex products in the marketplace are usually a combination of hardware and software. The products use quite complex hardware devices along with complex software to make a product.

Definition Phase

The product design process begins with a marketing requirements document that describes the requirements of the product. This document is designed to provide specifications to enable the product to successfully compete in the marketplace. This requirement is combined with other technological inputs to create an architectural specification for the product to be developed. This is described in the figure 3.1

This collection of requirements for the product are then broken down into specific requirements that can be implemented in software and hardware, keeping in view the long-term product goals and the current technology used to implement the product.

Subsequent analysis of the hardware requirements specification gives rise to the hardware architecture design process. The hardware architectural description is then created. The hardware architectural specification then is used to create a micro architectural specification. This specification is then considered as inputs into the verification planning phase.

Planning Phase

The verification planning phase is typically divided into a few phases. The initial planning phase typically gives rise to a verification plan document. The verification plan document describes at a high level, the overall approach to the verification challenge. The plan also describes the tools used and the strategies that will be deployed in the course of verification.

The high-level plan is then broken down into checkers and monitors and test cases and features in the environment as well as the testbench for the verification of the device. An example of a plan that is executable and track-able is shared at the end of this book in the section *Putting it all together*.

RTL development and verification

During the initial stages of the RTL development, the verification environment is also developed based on the specification of the device and the strategies adopted to verify the device.

The RTL and the verification are then run together and various tests are run on the RTL. During this phase, several bugs in the RTL and the verification environment are ironed out. The bugs that are found because of debugging are filed using the bug filing process that is described later on in this chapter.

As the RTL matures, the verification and the RTL are put through a regression process to ensure that the design is indeed making forward progress. Periodic reviews of the RTL and the verification code and the test cases ensure that the verification is indeed on track.

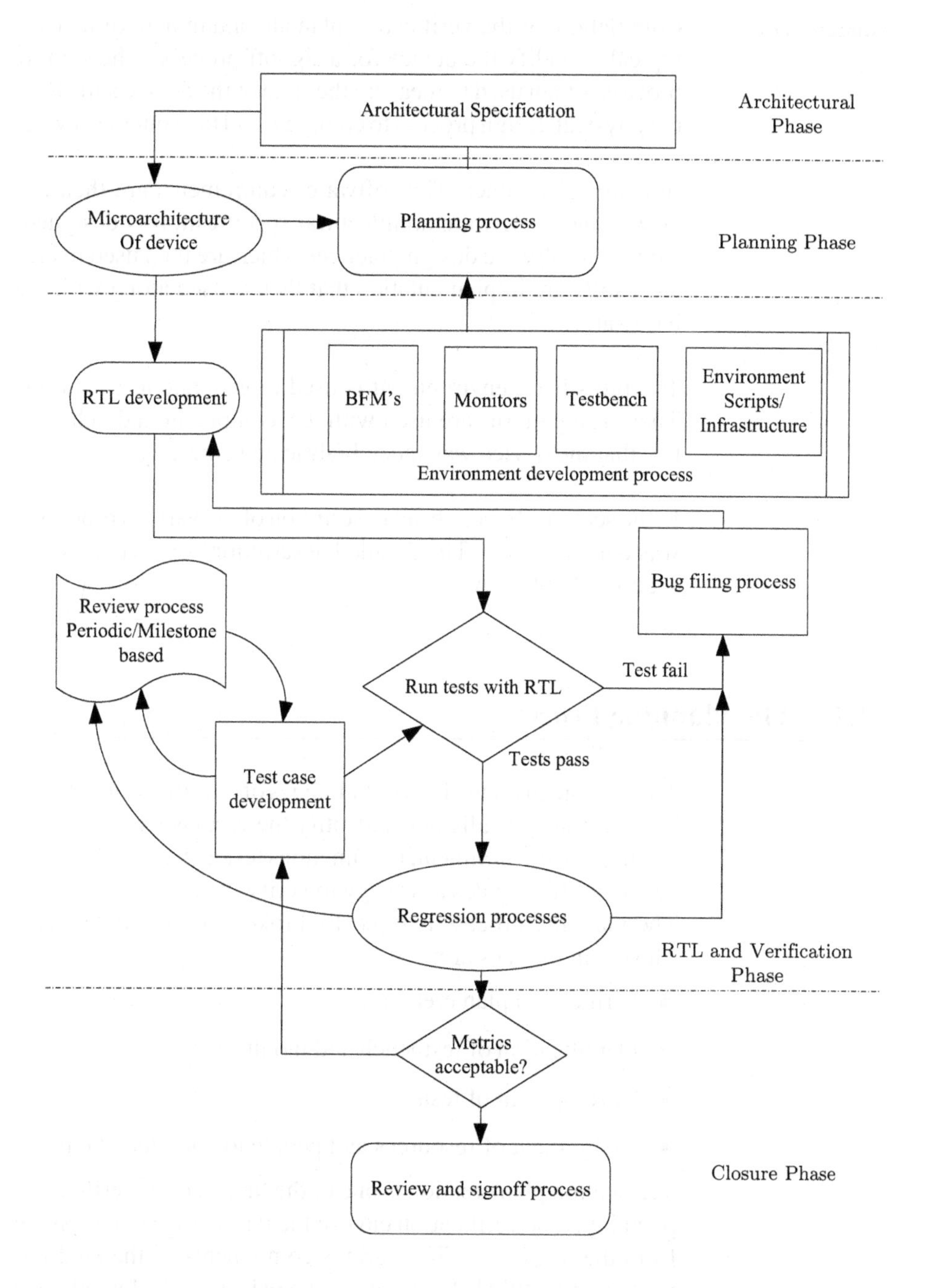

Figure 3.1. Overall View of the Verification Process

Closure Phase

Completion of the verification plan and attainment of metrics typically qualify the device for a signoff process. The signoff process is then used to measure the state of the device and allow the physical design process to complete and tape out the device.

In a similar manner, The software requirements are then analyzed and a software architectural specification is designed. Common software design practices which are then used to create a software implementation that then is used along with the hardware.

The pre-silicon environment is used to generate test patterns. These test patterns are used with test equipment and help certify that the device was indeed fabricated correctly.

In the section above, a brief description of the various processes was demonstrated. The detailed description is provided in the sections that follow.

3.2 The Planning Process

This is typically one of the initial stages of a verification activity. The process typically helps identify the various features in the verification environment, the functions tested in the various test cases etc. It provides a strategy to verify the device under test. The planning process is typically broken into multiple steps. The common ones are:

- Verification plan creation.
- Identification of testbench and monitors etc.
- Development of tests.
- Assignment of resources and people to complete the task.

Verification plan creation is one of the first steps of verification planning. The verification environment is distinct and separate from the test cases. The various components of the environment are identified. If a re-use strategy is indeed adopted, then the components that are being reused must be checked for compatibility and availability.

The test plan is then identified to define the various test cases. The plan scopes out the size of the verification activity and allows the planner to get a global view of the task. This test plan must be developed before any test cases are developed.

Before a test plan is written, it would be very useful to collect the following:

- An architectural specification.
- Different operating modes of the device.
- Kinds of behavior of the device in case of normal or erroneous input.
- Any specific standards that the device would adhere to.
- A list of I/O ports for the design.
- Interesting scenarios where the design might not behave normally!
- Applications where the device will be used etc.

The author has always found it useful to obtain feedback from the designers who know the internals of the DUV implementation and can help define interesting test cases that may be missed by the verification engineer. Hence, it is of importance that they should be involved in the process of defining the test plan.

3.2.1 Some Other Aspects of Verification Planning

Emulation and acceleration

In many designs , the software-hardware interface is verified using a hardware emulator. Hardware emulators offer the advantage of being able to run quick regressions of tests. Hardware emulation also needs some up-front work to be done to the tests and the environment before the tests can be run. Hence, some analysis must be done in the beginning to determine if any extra requirements imposed by hardware emulation must be considered. Many current generation devices now rely extensively on software. This factor must be taken into account as well and should be made part of the plan.

Gate simulations and regressions

In addition, the design may simulate well on a 32 bit machine in RTL simulations. However, the entire GATE netlist may be too large to fit on a 32 bit machine. Estimations of the design gate count are usually available early on. Consultations with the simulation tools vendors usually helps identify any potential concerns.

Gate regressions also are extremely slow and difficult to debug. Typically one of the main challenges is the size of the netlist along with X propogation issues that tend to consume a great deal of time if not planned for.

Mixed mode simulations

The device under verification may also be a mixed mode device. Many a time, there may be the use of other programs like MATLAB[1] that may be used. There may be the use of reference data files or associated programs which may need special licensing or large amount of compute power/disk space. The appropriate simulators may require special handling or considerations as well.

3.2.2 Verification Resource Planning

Resource planning is a vital step for a successful verification. The size of a verification task can predict the simulation hardware resources and the needed personnel.

Similarly, the number and complexity of IP's in an *SOC* or other complex device will determine the amount of estimated regression time, hardware computing resources, and simulation license requirements.

Software may also be a concern. Some specific software may run on operating systems like HP-UX or Solaris[2] or Linux. In addition to this complexity, the software may run only with some very specific patches installed and incompatible with the rest of the compute environment and shared among multiple projects.

[1] Matlab is a trademark of The Mathworks Inc

[2] trademarks: HP-UX belongs to Hewlett Packard, and Solaris belongs to Sun Microsystems

Schedule of large and expensive shared resources may also be a concern as evidenced in some larger companies. Emulators and related resources may be shared amongst multiple projects. The compute farm and associated queues may also be shared amongst multiple projects requiring detailed planning.

3.3 The Regression Process

A regression test suite is a collection of tests that can be run on the design in an easy manner with minimal effort. The main idea behind a regression stems from the ability to test everything when the design changes. Such an approach ensures that the design is always kept in a "Working state".

Is regression testing even important?

On the surface of things, a small tweak to the design may not indicate a extensive round of tests. The change may actually lead to a problem though. If the design is part of a larger system and some blocks were impacted by a change, a regression provides assurance that the change did not wind up breaking something else. Hence the quality of the design is maintained by ensuring that changes made do not break any functionality that was verified before the new change was introduced.

Tool and script versions also have their impact on a regression results. A newer version of a tool or script may behave differently from the version initially installed. Running a regression usually helps flush out any issues that can crop up later.

One important consideration is that regression tests should be repeatable and traceable. A test MUST be repeatable. If randomness is built into a test, then it is important to specify a SEED for a test in order to replicate an error detected on a previous run. Traceability is defined as having enough information to be able to debug the problem.

A graphical representation of this process is provided in the figure 3.2. Typically, in most organizations, the regression process is a periodic event that is usually triggered by some fixed parameters. For example, there might be a weekly run that is started at a specified time on a certain day of the week. In other cases, it might be when a significant RTL milestone is achieved.

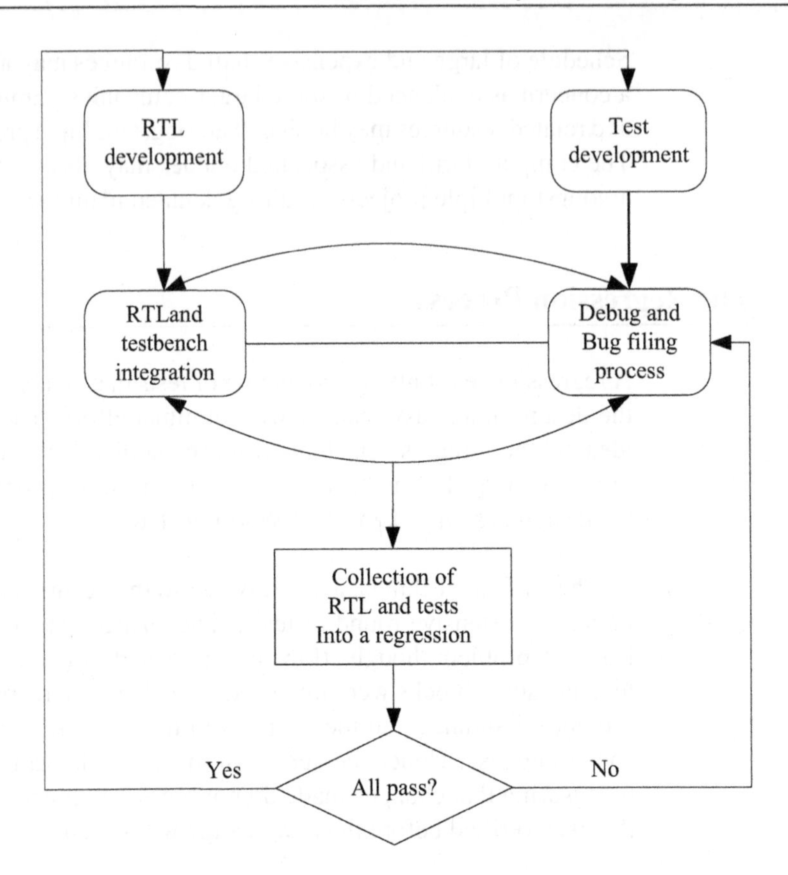

Figure 3.2. The Regression Process

One of the main factors that ought to be considered is that the regression must run as quickly as possible. The information available from this process usually signifies that nothing that was known to be working was broken when changes were made to either the tests or the environment or the RTL. (Except when running random regressions to explore states of the design). The number of machines and speed of the machines available to run the regression amongst other factors govern the speed at which the regression runs in this process.

Keeping the regression process going is an important task. Many organizations typically tend to assign an individual in the team to make sure that this process happens properly. In order to keep this process going smoothly, it is important to make sure that the various parameters like disk space, machine availability etc. are well taken care of before the jobs are launched.

Having to clean up a broken regression is no fun at all. Some organizations use a policy of aborting the regression as soon they can determine that the RTL under verification is indeed broken in a certain release.

Another important consideration to this process is that there must be some change to the RTL or the tests or the verification runs to warrant a new regression run. In many organizations, the regression at times takes many days to complete. Hence having a broken regression run implies that it will take many days to certify that the RTL passes the regression. This is in direct conflict with the fact that there should be small number of changes accepted before a regression is launched to enable debug and traceability. Usually what has been observed is that there is some sort of a compromise reached between the number of changes and the length of the regression based on the circumstances at that time.

Different organizations follow different strategies to verify their devices. Depending on the size of the modules being tested, the steps in the process and timelines vary.

3.3.1 Block Regressions

Block regressions start when the blocks or modules are available.

This book characterizes subchips (portions of a chip) and modules as blocks in this book. For purposes of discussion, this process is typically followed when there are a number of blocks that are verified independent of one another and have significant complexity and development time associated with them. Block regressions are typically characterized by having some bus functional models at their interfaces.

This process is typically followed when the block or module is reaching a state of maturity and a fair number of tests are developed. The module has many of the features expected of it implemented. The block level regressions help enforce a level of quality when subsequent changes are rolled in.

One of the key aspects of a block level regression is that most reviews typically expect that the module has attained 100% coverage metrics. It is difficult to get complete coverage, but a sincere attempt must no doubt be made.

There are typically no waivers granted except for pressing technical reasons for any statement or code coverage items. It must be noted that block level regressions happen early on in the development cycle. When the design matures, the block level regression activity usually tapers off and is replaced by the chip level regression process. At times, block level regression results may be combined with chip level regression results to present a more holistic view of the verification effort during a signoff review for the device.

3.3.2 Chip Level Regressions

The chip level regression is an extension of the block regression process, the difference being in the size of the design now placed in simulation and the types of tests being run. Frequently the tools and machines on which the regressions are run may be a little different from the block level regressions in case there are some capacity issues on the machines on which the simulations are run.

From a process angle, however, this is a similar process to the block regression process. However, it varies in the fact that the simulations are much longer and the environment is probably larger. Changes in the regressions are usually handled more carefully than in the block regression stages since the penalty for a failing regression is quite high.

Chip level regressions are usually used for the signoff review process. They typically start once the design has reached some point of maturity. The start of chip level regressions is usually marked as a significant milestone in any device development. *(read that as a celebration party !).*

3.3.3 Coverage in Regressions

Coverage can be classified into functional coverage and code coverage. Functional coverage focuses on the features in the design. Code coverage on the other hand measures whether various components in the RTL were indeed exercised by the regression. Coverage is nowadays usually available through a set of tools that are used to measure the degree to which the verification has exercised the design. The code coverage is typically available by instrumenting some specific monitors into the RTL and subsequently running the regression. Many a time, the coverage mechanisms are built into the simulator and can be invoked by using some specific simulator commands. In some organizations, functional coverage is measured as well. This coverage is measured using tests and specific infrastructure that is present in the environment.

Code coverage in regressions usually attracts some sort of a performance penalty because the simulator now has to do some additional "housekeeping" to report the various coverage metrics like line, code, statement coverage etc. Earlier coverage approaches used to "instrument" the code with special monitors and then run the coverage regression. Many organizations run coverage in regression on a periodic basis to see the progress of the design. (Some others do on every regression though!). During the final stages, the coverage from the regressions is usually reviewed to ensure that there are no missing test scenarios. Any exceptions are carefully scrutinized and then signed off before a device tapeout.

3.4 Maturing of the Design

In the chapter *Approaches to Verification* various approaches to verification were considered. Different approaches were deployed at either the block level environment, the system level environment and at higher levels of integration. The regression processes in this chapter described the various activities as well. As the design matures, the levels of integration of the modules increases. Different regressions are launched to ensure the health of the design. Initially the design startsout with module level regressions. When the blocks are mature,these regressions are completed and sub system regressions are predominant in the regression activity. In a similar fashion, the chip level regressions become predominant when chip level verification is taking place. This progression of the design toward maturity is shown in the figure 3.3. As indicated in the

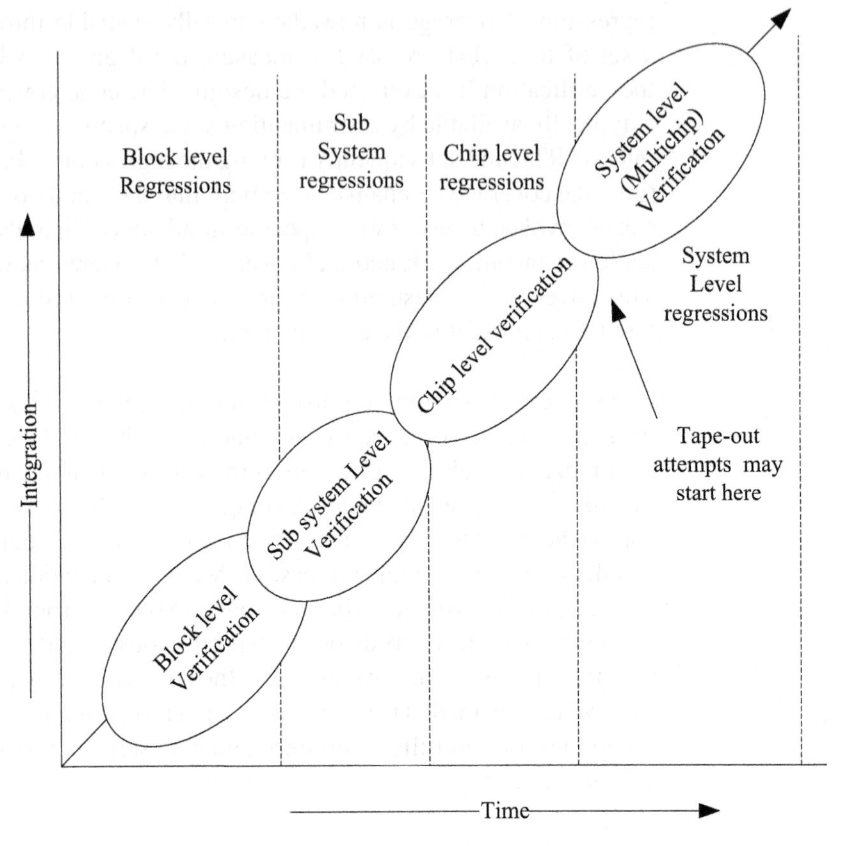

Figure 3.3. Levels of Integration

figure, the tapeout activity usually starts at the end of chip level regressions or system level regressions if applicable.

3.5 The Periodic Review Process

In this process, the various features and functions of the device are reviewed for understanding that the various tests actually cover the scenarios expected of the tests. Some of the typical questions that are asked during these reviews are:

- Do the tests cover all of the functionality required?
- Do we need any additional tests to cover the functionality of the device?
- Are there any additional random seeds needed?

- What are the holes in this approach to testing?
- Is there a issue with runtimes?
- Can we do this more efficiently?

Sometimes a detailed code review of the test code is also conducted to ensure that the test code does what it is supposed to do. This is similar to the RTL code reviews conducted during the RTL development process.

A code review is useful since a great deal of details typically emerge from the review. However, there are usually many tests in the regression. This makes a code review of the tests somewhat difficult to implement. Breaking the code reviews into small chunks helps make the entire process more manageable. Such an approach is presented in the *Putting it all together* chapter later on in the book.

One of the things to look for during the functional test reviews that helps identify problems is the fact that the code coverage (line, toggle, expression) for the block in question should be at 100%. Anything other than 100% is a dead giveaway that something is wrong somewhere. If there is a chunk of code that is not exercised by the tests, either the code is a chunk of dead code or the test suite is not complete, or something else is going on that warrants an investigation.

3.5.1 Regression Result Reviews

This process is used to review the regression results. It is usually informal in most companies in the initial stages. When an external customer or reviewer is involved, closer to a tape-out phase, the process involves going over the regression results, the tests and the functional features of the device to ensure that the device has indeed been tested to the best extent possible.

Feedback that arises out of this process sometimes results in either addition or modification of tests or the environment. This process is typically one of the many processes involved during the tape out signoff activity.

In many organizations, the regression results are reviewed regularly to make sure that the design is indeed making forward progress. If analysis of a collection of regression results indicates that the number of tests passing in each regression run is not increasing, then usually an investigation is launched to figure out where the problem lies.

3.6 The Verification Signoff Process

It is human nature to prefer binary answers to questions. Engineers being human, are no exception. As much as possible, engineers like to prove that something is either true or not.

Unfortunately, using such binary decisions in determining closure in chip verification is extremely challenging. For the most part, answering the question of whether the chip functions as the designer and architect intended is very hard, In some cases it is impossible to state that a design has been completely functionally verified.

The fact is, achieving true functional verification closure, and therefore taping out with full confidence, never happens. The decision to tape-out is always a judgment call. Typically, most successful engineers achieve a sufficient level of closure and confidence based on a combination of verification thoroughness metrics, the rate and complexity of functional bugs being found, and their own experience and judgment.

Towards the end of the design cycle, Verification signoff reviews are arranged when an ASIC verification effort has been more or less completed. During this review, various items are reviewed.

Typically, the verification sign-off process involves going over the test plan, device specifications, and the status of the tests. Under some circumstances, verification is incomplete for schedule or other reasons. Then care is taken to ensure that all high priority features are covered. Clear assessment of the risk of taping out is also made available. Some of the other items checked are results of gate regression as well as code coverage and functional coverage statistics. At this stage, any documentation issues are also addressed. During the review process, some of the items reviewed in addition to other metrics are:

- Number of checkers in the design
- Number of monitors in the design
- RTL Code coverage reports
- Specification stability
- Number of bugs found
- Vendor related checklists
- Test case density
- Test Object density
- RTL stability
- Bug find rates to look at the trend of bugs being found
- Bug saturation curves that reveal the trend of bugs found over a period of time

Special care is taken to ensure that all verification bugs are closed and that all checkers and monitors are indeed enabled for final certification. A detailed report of the features covered etc. is generated during this phase.

One of the other major factors that help make a decision are the bug find and closed curves as shown in figure 6.1 and 6.2 in the next chapter. The bug find rate curve typically rises rapidly and hits a peak and then begins to drop. A tape-out is typically attempted in the stages when it is apparent that there are no new bugs being found in the design in spite of efforts being expended over a period of time.

In many organizations, verification continues as a separate activity well beyond tape out. In some cases, the hardware emulation activity is also reviewed to make sure that there are no known issues at the time of the sign-off.

Verification sign off is a process that started with a great deal of care and deliberation since mistakes can cost significant amount of time and money and possibly market opportunity.

Conclusions

This chapter presented various workflows that occur during the verification cycle. As mentioned earlier, the specifics of the flow vary. It is hoped that people new to verification can use the material presented to familarize themselves with the various processes in verification. As mentioned earlier, there is a range of enforcement of these processes in various companies. This material should help the reader get well started as part of a ramp up to a career in verification.

PART II

INGREDIENTS OF SUCCESSFUL VERIFICATION

This part focusses on helping the reader become more successful in verification. As the reader is aware, success is made up of many factors. Some factors are abstract and some are well defined. There are three chapters in this part.

People make all the difference: People are one of the most important assets of any organization. This chapter presents factors and dynamics based on the author's experience that help teams and individuals become successful. There are two main themes to this chapter. The first theme is dedicated to exploring six qualities that help teams become successful. The second theme explores six habits that help an individual become successful in verification.

Doing it right the first time: This chapter explores various case studies from actual designs in verification. It discusses specific incidents where the benefit of hindsight indicates that things might have turned out better if things were done differently. There are seven small case studies in this chapter.

Tracking results that matter: This chapter provides an overview of various metrics that are used in verification. It discusses various metrics and provides an overview of the various items that are measured in the verification effort.

Chapter 4

PEOPLE MAKE ALL THE DIFFERENCE

Human Aspects of Verification

Verification is a team sport. Either the entire team wins or loses. There is no scope for any one person to be successful while others fail. Human aspects of verification play a great and important role in the way of execution of a project and its success.

Some of the material may seem out of context in a technical book on verification. However, all things considered, vis-à-vis how we do things is just as important as what we do with the verification effort. Small things ignored early on, have unfortunately become bigger problems later on. Successful teams have avoided some of the problems by using processes methodologies that take into account these aspects of verification.

Small companies and startups aiming to strike the marketplace quickly usually do not have a history of refining workflow processes since they usually attract people from other companies or assign roles to individuals in the company to accomplish the verification task.

The author had the privilege of working with many people from Silicon valley, and during the course of his career, the author found that most of the successful engineers whom the author had the privilege of meeting had actually developed a style of working. Largely, as the author learned from them, he realized that the following aspects were actually key to their success.

The author trusts that this chapter will inspire engineers and teams in their quest for verification excellence.

4.1 Team Dynamics and Team Habits for Success

Websters Dictionary[1] defines habits as

1. *A recurrent, often unconscious pattern of behavior that is acquired through frequent repetition.*
2. *An established disposition of the mind or character.*
3. *Customary manner or practice.*

A habit is usually understood to be a practice that is ingrained deeply in the conscience of a person so much so that it can be performed without conscious thought.

Team dynamics and people dynamics typically play an important role in the success of a project. Successful verification teams have a system of their own incorporated into defined workflow processes. These processes usually incorporate in some way or form the six habits incorporated described herein. Well formed teams typically comprise of members who complement each other's skills with their own and help the team succeed overall.

4.1.1 Habit 1: Begin With the Big Picture in Mind

Earlier chapters undertook discussion on the various aspects of verification. The cost of verification as indicated in 4.1 reveals the impact of a decision taken early on in the design cycle. Given that ASIC design is such an expensive process, it is imperative to ensure first time success on silicon

Stephen Covey [2] mentions in his book on 7 habits "*Begin with the end in mind.*" Verification is all about ensuring that the device operates the way it was supposed to and in a satisfactory manner.

In simple words, typically looking at the larger picture and attempting to solve a problem usually yields very effective results. This is one of the cornerstones of a successful verification strategy. Many of the readers have no doubt been in a situation where they certainly wished that they had done some thing or other differently in order to overcome some problem or other. Having the bigger picture in mind possibly could result in lesser work than was originally envisaged.

One of the best (and probably most common) ways to design an environment is to design it 'top down' and then implement it bottom-up. This approach ensures that all the parameters required to implement the verification environment and tests are indeed considered completely. Some of the advantages of doing so are:

1. The parameters are well defined. A sense of clarity and stability is imparted to the entire verification process.
2. The design and architecture of the environment is usually able to handle change very easily.
3. The end goal is always in view. The ultimate goal is what matters in the end.
4. The task can easily be broken into smaller pieces (see habit #3).

This habit usually is carried over all the way through from verification inception to tape-out. This habit needs to manifest itself into many forms.

Many a time, We have noticed that some or the other short cuts have been taken with consequences that have a negative effect (see the chapter *Doing it right the first time*). These consequences are not usually apparent when the decision is taken, however, in hindsight, the lesson is usually apparent.

Consistency in adopting this habit is important to the success of the team. This habit also provides rich dividends in terms of time saved. In one instance, a verification team building a very complex ASIC realized that it would be a real challenge to write several tests for pre/post silicon in the time frame involved. The team then proceeded to identify the manner in which the real device would be tested. The information so obtained was then used to architect a verification environment where the tests from the pre-silicon environment could be used with very little modification to validate the ASIC once it returned from the ASIC foundry. The environment also included features to enable tests written at the block level to be easily reused at the top level. The implementation of this habit alone saved the team from writing and working on several hundred tests.

4.1.2 Habit 2: Do it Right the First Time

In simple words, think it through and do it once after all the parameters have been considered. The task must be undertaken in such a way that there must be no need to visit the topic again and all eventualities and conditions are addressed in one attempt. In all probability, this approach will wind up with the most optimized solution for the particular situation.

A decision taken early on, in the process can have a remarkable impact on the entire verification effort. In many cases, procrastination could be sometimes quite dangerous! For instance: One may choose to verify only a portion of a module without considering completely if the coding styles or guidelines are required for top level activity, there is a chance that the work may need to be redone again.

As an illustration, the events that happen when a decision is postponed usually leads to:

1. Quite a bit of work is probably expended getting to a point where the decision to postpone has already been taken. Some of the effort is wasted since it did not go all the way to conclusion.

2. Someone now needs to track the item down and ensure completion.

3. When the item is revisited, there is a reasonably a good chance some time will be lost recollecting what was done and what was not.

4. Sometimes the details of the issue under discussion are lost.

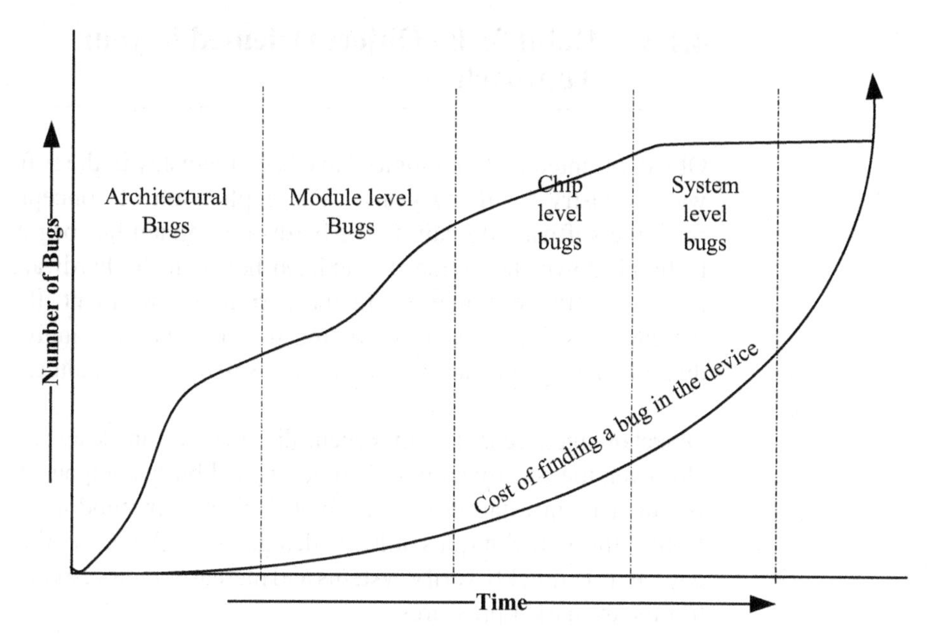

Figure 4.1. Cost of Finding a Bug in an ASIC

As the reader can observe, a simple *'I will take care of that later/I do not think I care about that right now'* can balloon into doing a fair bit of work all over again. This is a fact. Readers will recognize that most of the time, any deferrals in decision making invokes Murphy's law[1] and nothing ever happens the way it is supposed to.

The figure 4.1 illustrates the cost of finding a bug in a design is very small in the initial stages. This is usually the case since the turnaround time for fixing a bug is very small. A bug found in a block level test bench could be fixed very quickly since the complexity and run-times are not very high. However if the same bug is deferred to the sub - system level, it becomes a little harder to find the bug. It may also be a little more time consuming. If the design is in the physical design phase, the cost of fixing the bug increases, since an extensive regression is needed to actually validate that the change did not break anything else.

[1]This law says "'If anything can go wrong, it will"'

4.1.3 Habit 3: Be Object Oriented in your Approach

Object oriented methodologies have been mainstay in the software industry for many years. The application of concepts from the software domain to the hardware domain has begun to break down the divide that existed between the hardware and the software domains. In the current environment, the increased use of software modeling practices has led individuals modeling software appear more like software programmers.

Object oriented techniques may seem alien to hardware designers. However, the construction of modules using library components is but a manifestation of the object oriented methodology. Using these techniques allows designers and verification engineers to quickly verify systems and increases their response to time to market pressures.

Using object oriented techniques in itself have many advantages as evidenced in, [3], [4]. The advantages of these techniques for hardware design include:

- Models can be maintained and reused extensively;
- The ability to alter general-purpose components to more specialized components;
- The possibility of employing existing software synthesis and verification techniques.

4.1.4 Habit 4: Reduce, Reuse and Recycle

Reuse is all about trust. It is also all about leveraging existing work that has already been done.

Reuse in verification reuse implies reusing the existing verification environments or its components. Components of verification environments developed for the other designs or blocks or previous versions of the design are usually reused. These components are typically the monitors, checkers, tests and scripts, BFMs etc that comprise of the environment.

Given that the previous habit discusses object oriented practices, does reducing and recycling even merit being called a habit?

Many a time, object oriented practices may not be followed. This may be particularly true of older designs. However, it is definitely possible to reuse portions of the effort expended when verifying the older design to save work.

The complexity of design verification is exponentially greater than the design complexity. Verification today consumes around 60-80% of the overall effort. Reuse can bring the effort down to a smaller number.
Reuse has some distinct advantages: it can reduce the development effort and reduce the risk from a verification point of view. In some cases, it helps shield the verification team from having extensive knowledge about the modules that are being reused if the modules are proven in silicon and are not being altered.

In order that reuse is practiced as a habit, the modules must be engineered with reuse in mind. Some of the typical considerations are:

- Documentation.
- Ability to be extended and altered.
- Customizable based on the environment.
- Easy to use.

Another discussion and some good practices for reuse and their implementation are found in [3] and [5].

4.1.5 Habit 5: Innovate

Change is the only permanent thing in this world

Designs usually increase in complexity and features. Doing things the same old way no doubt offers the opportunity to tread the beaten path, but given the change that is being implemented in designs today, exploring new methodologies and techniques is essential. In fact, it becomes a mode of survival for verification teams given the changing technology landscape!

In many projects, after the tape-out phase, a period of introspection usually leads to a set of lessons learnt as a result of the ASIC activity. These lessons are important since they have been earned with the hard effort of experience.

An example: In one version of a chip the author worked on, the register map for the device kept changing due to the change is the specifications of the device. Unfortunately, the design engineer was not able to contain the changes since many matters were not in his control. This led to the tests being reworked a total of five times through the life of the project! There were 3 different versions of the tests at one point in time with some tests being ported from one version to the other just before a regression run. The design engineer came up with a script that helped the tests to migrate easily, but this could have been avoided with a different coding style.

In the next device that the author had the opportunity to be involved with, the first change that the author made was to move the register map to a C++ based environment where the changes were isolated to a database and a header file. The rest of the verification environment was actually derived from the database and these files. As the design moved, all that was done was to update a single source and the rest of the environment was automatically up to date when the RTL had changed to accommodate the different specifications. It turned out that there were 17 changes in some register definitions over a 2 month period!

Not having made the single change based on prior experience would have been disastrous in this project. During the course of implementing the change, the author did encounter some interesting viewpoints. However, the results paid off later on.

4.1.6 Habit 6: Communicate

This habit is one of the most important habits that a successful verification team has had. Communication takes many forms. Some of them are:

The grapevine – I told you about that bug!

The grapevine usually exists as conversations over the cube walls. For the most part, this seems to work in small well-knit teams. It seems to be the case for many of the module level verification teams who share information constantly back and forth. For the most part, as the size of the team increases, the approach becomes impractical. Consider a module with multiple verification stakeholders. Any stakeholder who was probably affected by the conversation and not in it misses out on learning about the issue. In addition, the possibility that the person who was not part of the conversation will find the exact same bug is very high.

The other problem with this approach is that there is no record of what was found and fixed. There is no way to show that the number of bugs found and the bug find rates are tapering off.

The email system- I sent you email.

This has a wider audience than the first one. Many projects in many time zones are sometimes run this way. Email is no doubt one of the great ways of communication. However, people are invariably inundated with email. There is the possibility that the particular bug was overlooked. In addition some of the verification stakeholders might not be copied on the email. Somewhere along the line a "data retention policy" might kick in and some vital information that should have been tracked may be lost.

A team meeting

Unfortunately, this seems to work well in a small gathering. The presentation is shared with people in many places as a communication medium. What the author has liked about this medium is that there is the opportunity to interact with others in a discussion and get some more information.

People are however hard pressed to keep taking notes in these meetings and some essential items have the possibility of not being assigned the priority required. A significant amount of time is wasted and the meetings are expensive from an engineering point of view. In addition, there is the possibility that some important stakeholders may be missing from the meeting for a variety of reasons.

The bug tracking system

This approach typically involves the use of a central database with a web based or graphical user interface of some sort. Some organizations use a spreadsheet, which may be centrally maintained. The main advantage of such systems is that the common problems that are listed above are completely avoided and there is a permanent record of the issues found in the design. Various metrics are enabled by use of such systems that have become very popular today. Various companies provide a variety of systems with sophisticated features to enable their customers to track bugs and changes effectively. The discussion of these is beyond the scope of what is presented here.

No doubt, other methods not presented exist. In addition, It is noticed that verification teams are closely in harmony when a team member shares the information about a problem he/she is facing with the rest of the group rather than allowing other team members to encounter over the same hurdles again. Many a time, the author have seen an alternative approach to the problem coming up with a more optimized solution than was originally envisaged.

4.2 The Six Qualities of Successful Verification Engineers

The various habits of a successful verification team are described in the previous section. It is apparent from the discussion that team dynamics have a profound effect on the quality of verification as a team.

On the other hand, in addition to team habits, how an individual approaches a challenge is also an contributing factor to a successful tape-out. In that context, the sections below explore the various individual qualities that help a verification engineer become successful in addressing the challenges at hand.

Websters dictionary[1] defines qualities as:

1. *An inherent or distinguishing characteristic; a property.*
2. *A personal trait, especially a character trait:*
3. *Essential character; nature:*
4. *Superiority of kind:*
5. *Degree or grade of excellence:*

As the reader will discover, all engineers possess the qualities described below in some shape or form and that the qualities are essential to success.

4.2.1 Quality 1: The Ability to see the Full Picture

The charter of verification is to verify that the design is operating correctly as per the specification of the device. Consequently, the verification engineer's view of the problem is quite different from the designer's view of the problem. Challenges abound in verification since the amount of work is larger and require many techniques to address problems. This implies an ability to get an overall picture, which encompasses the design and the surrounding environment.

The scope of verification includes:

- Being able to understand the parameters of the device itself.
- An ability to specify the items that need to be tested.
- The ability to interpret failures.
- Identify simulation bottlenecks and isolate them.
- Thoroughness in resolution of various issues that crop up during the verification phase.
- Verification is indeed a costly business. This has been evidenced earlier. Any contributions towards cost cutting are always welcome in any organization.

Illustrating the final point above, consider an device being tested that dumps out the data and uses a pre-post processing approach to verify device behavior. If the verification team chooses to run the simulation along with the post-processor in parallel, not only will the simulation slow down, it may cost more depending on the number of licenses for the post processing tool. Using some sort of a queuing mechanism may help efficient post processing as well and save the organization money.

Debugging a problem with a failure in simulation usually involves looking at various factors. The manifestation of seeing the full picture as a quality typically helps the verification engineer to narrow down the source of the problem before significant time is lost in the debug effort.

4.2.2 Quality 2: Assumes Nothing

One of the important aspects of verification is that no assumptions are made about anything in the design or the architecture of the device.

Making an assumption about something or the other is the death knell for verification. Making an assumption can actually limit the space of verification and mask out some important bugs. In addition, the assumption may be invalid which may be counter to the original charter of verification itself!

In many instances, the design and architecture teams may have made some assumptions of the system which may be proven to be incorrect by verification.

The author does not suggest that it is possible to eliminate assumptions completely. Assumptions can no doubt be made, but it is absolutely crucial to document the assumptions that have been made. It is also important that the assumptions made are made out to be facts prior to the tape-out of the device.

4.2.3 Quality 3: Consistent

Webster's dictionary defines consistency as

> *n. Reliability or uniformity of successive results or event.*

Consistency is one of the more important habits of a verification engineer. It reflects the fact that the verification engineer can be relied upon to perform his task. Consistency breeds repeatability and reduced confusion. Being consistent implies that the results so obtained can be relied upon at any time.
Many a time, one typically observes the following sort of conversations: *(the author is sure the reader has certainly observed similar ones in his career!)*

- Why do all these messages look different?
- Why does this not cause the test to fail?
- Do I have to do this all over again?

- Can we not get more of this like the other test that just ran?
- How come this does not look like the other tests ?

Consistency helps avoid such concerns and issues. If the tests and the environment use a standard coding style that is common and pre-planned, then the benefits of having simpler implementations which are easier to debug and maintain.

4.2.4 Quality 4: Organized

Typically, design verification begins with unit level verification wherein there are large numbers of signals and many combinations to be tested. Keeping track of all these can be an onerous task as SOC designs tend to get larger and larger. The number of features to be verified increases dramatically. Being organized is one of the key attributes to help keep the complexity at bay and allow task accomplishment on schedule. An object oriented methodology can itself help in introducing a sense of structure throughout the project. Since many bugs also do tend to manifest themselves during this process, the ability to see the root cause and identify the symptoms of existing identified bugs.

4.2.5 Quality 5: Multi-skilled

Todays environments are complex. They are a homogeneous integration of a variety of software, operating systems and languages. The successful verification engineer would have a variety of languages in his repertoire. The verification engineer typically has to be able to write scripts and analyze designs. In addition, to find a problem, the engineer typically has to be able to traverse many levels of abstraction. The current verification environments today require a fair amount of sophistication to use them.

Being in possession of multiple skills is now a must in verification. It is no longer sufficient to know a particular type of approach or HVL in order to be successful at verification. This quality also signifies the ability of the verification engineer to rise and adapt to the challenge posed by the current crop of complex devices.

4.2.6 Quality 6: Empower Others

Todays SOC challenges are large and complex. The current designs typically require the effort of many people to ensure that they are bug free. Working in a team implies that the verification engineer is able to work with other design groups and able to adapt to their terminology as well. This in other words implies the ability to synergize with other team members.

The American heritage dictionary [5] describes synergy as:

> The interaction of two or more agents or forces so that their combined effect is greater than the sum of their individual effects.

Many verification engineers typically come together to work on a ASIC verification effort. In a typical situation, each engineer brings a bag of tricks and experiences to the table. Typical teams have a diverse set of skill sets which are all required to complete the task correctly.

Since the scope of design verification is vast, Synergy within the group typically leads to co-operation between team members. This synergy can take various forms ranging from assisting team members challenged by a certain problem to sharing best practices for success.

Being able to recognize each team members strengths and leverage them for the project success is vital to every verification engineer.

Synergy has many positive effects. These include an increased level of confidence and energy in the members of the group as they progress toward device tape-out.

Conclusions

This chapter has explored various human factors that have made or broken teams in verification. Six habits practiced by successful verification teams and the six qualities of successful verification engineers were explored. Human factors, though hidden for the most part, can play a crucial role in the overall success of a project. Depending on the team's synergy, human factors have the power to make an enormous difference in the quality of verification. Hence, the human element of the verification process is not one that can be overlooked.

There are no doubt other important lessons that can be learnt by studying how human factors affect ASIC development and execution. Over a course of time, one of the main lessons that the author has learnt is that successful teamwork and the practice of the habits above are strong contributing factors to a successful tape-out.

References and Additional reading

[1] The Houghton Mifflin Company (2000). *American Heritage Dictionary of the English language*. Fourth edition.

[2] Covey, Steven (1990). *Seven habits of highly effective people*. Seven habits of highly effective people. Free Press, New York, NY, USA.

[3] Yee, Steve (2004). *Best Practices for a Reusable Verification Environment*. www.design-reuse.com, New York, NY, USA.

[4] Kuhn, T., Oppold, T., Winterholer, M., Rosenstiel, W., Edwards, Marc, and Kashai, Yaron (2001b). A framework for object oriented hardware specification, verification, and synthesis. In *DAC '01: Proceedings of the 38th conference on Design automation*, pages 413–418, New York, NY, USA. ACM Press.

[5] Albin, Ken (2001). Nuts and bolts of core and SoC verification. In *DAC '01: Proceedings of the 38th conference on Design automation*, pages 249–252, New York, NY, USA. ACM Press.

Chapter 5

DOING IT RIGHT THE FIRST TIME

Case Studies from the Real World

In this chapter, a few case studies covering some issues found in real life verification environments are presented. As a result of the case study, it is hoped to learn from the past and use the information herein to better equip us all in the future.

Writing a verification plan is no mean task. Designing a verification environment is even more so difficult. However, in the current day and age, some thought must be devoted to how the entire verification plan is prepared. Designs are getting increasingly complex and the time available to turn around the design into physical silicon from concept to product is decreasing. This unfortunately means that there is a double whammy on the poor verification engineer. The case studies below suggest things that the reader might consider as thoughts before designing the environment for verification.

For purposes of the study, The names of the company where the incidents occurred and the specifics of the parties involved in the incident have been masked for reasons of confidentiality. Some of these might sound extreme, but truth is sometimes harder to accept than fiction and have actually happened!

5.1 Block and System Level Tests use Unrelated Environments

This particular case was a complex control block that acted as an arbiter. The entire environment at the block level was written in *e*. Quite a few monitors and checkers were actually instantiated in the block level environment. The top level was written using a different approach. Consequently there was very little re-use between the block level verification and the system level!

An analysis of this approach revealed several problems. First, the verification engineer had spent quite a bit of time to test the block level. The tests were not reusable because the coding style used poked actual signals in the RTL. In addition, tests had to be rewritten at the top level all over again. When a bug was found at the top level, the poor engineer had to go through quite a bit of trouble to recreate the tests in the block level environment thereby doubling his effort to verify the same block – Time that would have been better spent elsewhere!

An additional problem that arose from this approach was that keeping track of issues overall became an issue. Not only was the schedule delayed overall because of the increased verification effort, but also effectively increased the overall cost since several things like salaries for people involved, tool costs, missed opportunities came into play.

Conclusion: *Attempt to reuse everything possible rather than reinvent the wheel.*

5.2 Not Implementing Monitors and Assertions Early on

The author has actually been in companies where the monitor development was the least of the priorities in the verification process. Interestingly, he has noticed that people think that they should get all the tests written first before getting the monitors or assertions done. More often than not, they never were done. Unfortunately, this practice causes quite a bit of grief overall in terms of the uncertainty that arises.

This particular case was of a memory controller block, which had been verified with a great deal of effort. The block was passing all the tests defined so far and a case was presented to sign off on the verification of the block so that the engineers could be moved to other tasks. During the review, it was noticed that many of the monitors were not implemented completely. Consequently, it was decided to implement monitors into the verification environment so that some statistical data could be collected prior to a formal sign-off.

Interestingly the coverage monitors revealed that only 40% of the features were actually exercised by the verification environment. Additional tests were needed to fill in the rest of the gap! The closure process took an additional 4 weeks overall to declare the block as done.

Analysis revealed that this information would have been available beforehand if only the monitors were instrumented in the code. Tests could have been more efficiently written and time spent a little better doing other things that mattered.

Several studies that have been done actually prove that adding assertions and monitors *actually reduced time instead of adding to the burden of verification.* The thought to consider for all who propose that this is wrong: *the time it actually takes to write a monitor is much less than the time it actually takes than the time it takes to debug something without the monitor put in.*

Conclusion: *Write the monitors and assertions early on. You will learn more about the design up front and save everyone a lot of trouble overall.*

5.3 Review Processes Not Done Timely

The author has actually had the benefit of seeing quite a bit of this error in quite a few companies! Unfortunately, this has become quite common since the common excuse is that there is no time to do it!

This was the case of a network processor block, which was designed and verified by four junior verification engineers and a couple of design engineers. They were working on the block for a couple of months and whenever they were asked for a review, the reply given was that they did not have the time to get it done.

Finally, when a review was conducted, it was revealed that there were some things missing in the overall strategy. Many of the tests were non-portable by nature of the constructs used to build the tests. Some of the implementation needed a little rework. By this time, a couple of months had passed and it was probably going to take an additional two months to get it all done.

Analysis revealed that had the reviews been conducted on time, this could have possibly been averted somewhat. The other side to this episode was that another senior engineer found a quick way to salvage quite a bit of work and it only took an additional three weeks get it close where the group wanted the verification effort to be.

Conclusion: *Review work regularly. This helps keep everyone on the same page and avoids surprises later on.*

5.4 Pure Random Testing Without Directed Tests

There have been many debates about the virtues of directed tests vs. random tests in a variety of verification circles. The author is of the opinion that each has its own place in the verification strategy.

This particular scenario was a state machine, which was verified using the high-level language *e*. The test writer had written all the tests as random tests without developing any directed tests. All things appeared fine on the surface although there was a hitch!. There was not a single regression where one or the other seeds failed. The tests were getting longer and longer. Some ran a good 24 hours before failing. All this was giving a great deal of anxiety to the project leads and the manager of the group as no end was in sight.

Analysis of the specifics of the problem showed that a few directed tests could probably hit the same scenarios that were being looked for in random testing. Not enough seeds were run in the first place. Proper metrics on bug closure were not applied as well. Doing so would probably increase the level of confidence in the design to a point where decisions could be taken on a course of action.

In another conversation the author had with the design manager of a large company, the latter revealed that using randomness, they were able to hit 70-80% of the coverage points quite easily. However, the last 20% of the coverage points was an uphill battle taking a long time. It is the author's opinion that a decision on using directed tests or directed random to help cover the rest of the coverage space would have probably yielded a greater return on investments.

Conclusion: *Use random testing wisely. It is good for some things. Some things are better served by directed tests. An analysis of the situation will reveal which approach will help get the biggest bang for the effort invested in verifying the device under test.*

5.5 Not Running a Smoke Test Before a Regression

This has been a source of battles for as long as the author can remember. Unfortunately, this has a huge impact on the overall verification. It is one of the big time-wasters on the author's list.

SMOKE is a test that checks the design for sanity after making changes to the RTL. It ensures that a certain minimum quality is maintained by the RTL before a regression begins. Unfortunately, too many designers overlook this. As a result, a regression run is launched and then several members of the verification team lose precious time going over the failures repeatedly. The author has actually seen instances of code that does not even pass the "Compile" test being checked in!

Unfortunately, what is not bought out is the amount of trouble this brings the verification engineer. Numerous CPU's are locked up, disk spaces to be managed, results to be parsed, explanations to be given and time lost just because a quality step was not met!

Conclusion: *Use a screen set of tests as a pre-acceptance criterion. It does not make sense to spend a whole lot of work especially if the basics are not working. One can always reduce those tests from a regression list if runtime is a serious issue.*

5.6 Lint Policies

The author actually had the privilege of being involved with a design that had over 20,000 warnings! Some of them were serious. The designer was insistent that the tests be written and run against the design before he fixed the lint warnings and errors. Unfortunately, the verification engineer was a junior person and a contractor to boot. Hoping to please all, the verification engineer set about to do his job. The bugs he uncovered were the same ones that were shown in the lint report. The difference being that the results were several days late and had a test case to prove the problem!

The test cases were redundant somewhat when the problem was fixed. A lot of time was wasted for no reason if the RTL was lint clean. No additional information was actually obtained outside of running the lint report and a considerable amount of time and money was wasted. The verification engineer would have probably been better off writing tests to exercise the design harder.

Conclusion: *Use lint wisely. It can actually save you a great deal of trouble by telling you up front where the problem might be rather than finding it the hard way.*

5.7 Effective Use of a Source Control Strategy

This case study shows an interesting aspect of the human trait that plagues us all! In a networking company developing an ASIC, some of the verification engineers were quite new to the job and had not completely understood the ramifications of using a source code control system.

The design was made up of several large sub blocks. The design in question was a large module. A designer and a verification engineer were paired together to ensure that the module was operational as is typical in ASIC development. There was a single small hitch. The designer did not want to check in code until it was bug free! He did not publicly announce his intentions though or he would have probably been taken to task (or crucified!) by all the people around him. The poor verification engineer had to work with the designer under these circumstances!

Being new to the job, the verification engineer kept quiet about this restriction posed on him by the designer. He would set out to verify the RTL and every time a bug was discovered, the designer would very quickly identify the problem and fix it. He would then hand the verification engineer fixed RTL to verify. No bugs were filed against the block since the version number of the file never changed until the module was verified! This went on for some time and what the rest of the team externally observed was that the version number of the file was not changing and the verification engineer seemed to be working very hard with no apparent output! There was no profile available as to the kinds of bugs that were found in the module because the team never wrote anything down.

Visibility into the module by other team members was terrible since it was almost non-existent. There was no profile that also tracked the development of the verification environment or the RTL either. None of the normal metrics that were used to measure RTL quality was available during the review.

The matter was resolved when the verification team finally insisted on checking in all the RTL code that was received and then verifying it. All the bugs were then filed on the version of the file that was received for verification. Interestingly this happened close to the module freeze and most of the information mentioned above was unavailable.

Conclusion: *Ensure that RTL that is verified is checked into a source control system. Verification should never start unless the module is completely checked into a Source control system. Doing so ensures that one can accurately reproduce bugs with a certain version of the file and ensure quality of RTL and the tests.*

Conclusions

This chapter presented various case studies in verification. As always, the benefit of hindsight is never available when the verification activity is in progress. While it is noted that many of the incidents mentioned above do not happen frequently in many modern environments, the learnings presented have been affirmed by many in the industry as common issues frequently seen in verification. It is hoped that this chapter allows the verification engineers to use the benefit of hindsight offered from the experience of many engineers to better optimize their verification efforts.

Chapter 6

TRACKING RESULTS THAT MATTER

Metrics In an Verification Environment

6.1 Why Do We Ever Need any Verification Metrics?

A manager's dream is to have complete insight into the design at all times

Execution of verification plans is something that all teams and organizations care about. In order to ensure that the plan execution is indeed on track, we need to look at metrics that define how well the verification effort is going. These metrics help communicate the state of the verification effort.

Janick's Book [1] suggests "*Managers love metrics. They usually like to attach a name and a number to various items that they are keeping track of, and assign a measure to track completion of the items* ".

Metrics are of all kinds. A large number of them influence decisions on the verification effort. Hence, accurate reporting and measurement of metrics is crucial to ensuring success in verification.

In this section, various metrics that are in common use in the verification environment are presented. These metrics are in common use in the industry to date and are used to guage the progress and performance of the design. The exact terminologies of some of the metrics vary from one organization to another with the core principles behind these measurements originating from a common source.

Metrics are a communication tool

One of the main advantages of having standardized metrics is that the metrics allow the clear and consistent communication of the design status to others who may not be completely familiar with the details of the design per-se. By merely reviewing the metrics, and having some knowledge of the design, many people can get a sense of where the design is.

6.2 Metrics in a Regression

Time duration

In a typical verification environment, regressions are usually run on a periodic basis. The regressions typically range anywhere from a few minutes to several days. In a scenario where the regression runs for a long time, it becomes of vital interest to determine quickly whether the regression has a problem or not (fact is, that is the only thing of interest!)

Several parameters affect a regression run. The time taken to completely run a regression depends on several factors. A list of parameters is presented herein. It is not an all-inclusive list in any sense of the word though.

A typical regression run usually has many tests to be run. These tests can be run in a variety of ways:

Running them sequentially on the same computer is usually acceptable when the total time of the regression run is small. However, when the overall time required to run a regression is large, then running different jobs on different machines usually gives an improvement in performance.

Running tests in parallel on multiple machines can be accomplished in a variety of ways, ie. dividing the jobs manually and running them on different machines or automatically doing so using a program.

Manually dividing the tests works fine for a small number of tests with some specific parameters. It does break down quickly as the number of tests increase. Using a software program to do the distribution of tests to multiple machines is usually preferred since it makes a repetitive job a whole lot easier. These programs are known as queuing or load balancing programs.

There are several well known programs[1] which do the job quite nicely. The use of queuing programs usually helps increase the machine utilization time quite significantly. This has been proven in several studies.

Effect of machine types

In recent days, this has become a fairly interesting topic. Earlier, design was mostly accomplished using large expensive Unix machines. With the advent of powerful desktops with clock speeds of over 3 G Hz along with operating systems like LINUX [2] makes it possible to run the regressions very quickly.

Why is the type of machine such a big concern? usually if one were to look at the code profile generated by profiling a simulation of a test, it is typically observed that the vast majority of instructions executed are arithmetic/logical in nature. In a typical computer, many of them are executed on a single clock/instruction. Hence, having a faster clock implies that the simulation runs faster. The higher the clock speed, the faster the test runs!

Effect of PLI/FLI calls

This is sometimes an overlooked parameter. Usually, using a PLI/FLI call in any environment is necessary. What is important though, is to keep the usage of these calls to a minimum and to use this feature in the environment cautiously.

Usually adding PLI/FLI introduces an overhead to the simulator. The overhead is typically of the order of 10-20 % just for using the call alone. Heavy usage of the PLI/FLI features typically reduces the performance of the simulator drastically. Time that could be used to simulate the circuit is now used in the PLI/FLI calls. This means that at any given time, there are less cycles passing through the device under verification.

Here is an example: Assume that a simulation runs for 100 hours. The PLI calls as measured account for 40% of the time the simulation is running. This implies that 60 % of the time, the design was actually exercised with some data passing through.

[1] LSF from www.platform.com, SGE from gridengine.sunsource.net, Queue from the GNU foundation etc, to name a few.

Assuming that x clock cycles passed through the design during the 100 hours, it is noticed that a reduction of the PLI Calls to 20%, would enable the design to run at least 20% more clock cycles through the design. This in turn could possibly stress the design a little harder leading to more confidence overall.

The author is not saying that it is possible to eliminate PLI/FLI calls altogether. Many modern languages (Vera [9], or *e*do use the PLI/FLI interface to communicate with the simulator. However, the benefits of using the PLI/FLI interface must be weighed against the overall penalty of the regression runtimes. A careful determination of what can be put in the PLI as opposed to placing it in a HDL testbench will definitely help the user optimize the regression runtimes. There are no doubt other methods like assertions from the Open Verification Library [11], The PSL language or other methods native to the simulator that could possibly be an alternative to coding checks that use the PLI/FLI to speed up the simulation overall.

Test Run Rates

This is another metric that specifies the rate of tests running over thenetwork. This typically is a representative number of how quickly the regression can be done. This parameter usually is affected by the number of machines available to run a regression and the software used to run the regression

Data Dump Policy

It is a well-known fact that the more the data one has to log and storeon the disk, the slower the simulation will get. In a typical regression, most users would like to obtain the results of the regression quickly. This implies that the amount of data being written to disk has to be kept at a minimum (the minimum being the data required to debug the test effectively in case of a failure).

Tip:*Turn off the use of a waveform dump in a regression. This will slow down the regression and if the test passes, provide possibly no useful information at all. Use specially designed monitors if you would like to keep track of specific signals instead.*

6.3 Commonly used Metrics

Number of checkers and monitors

The number of checkers and monitors in a design are a rough indicator of the verification environment implementation. It is usually implied that a large highly complex design should have a sufficient number of checkers and monitors.

Is there a real problem if there are not enough checkers and monitors in the environment? This is probably not true in many cases. The major problem experienced if one does not instrument enough monitors in the environment is that there is lack of visibility into the design. In case there is the need to debug a problem, the issue at hand becomes a time consuming one.

Specification stability

This is indicative of the number of specification changes in the design. Changes in a specification are indicative of the number of changes in the design, leading to changes in the verification environment as well. In the ideal case, a specification for the device is created and then the device is designed. Unfortunately, that is never the case.

Due to either technical challenges or other reasons, the specification may change. When this happens, both the design and verification are impacted. Assessing the changes and responding quickly to the changes in specification is frequently crucial to the success of the device in the marketplace.

Using a methodology that encapsulates in an object oriented manner, various functions in the functional specification is essential to contain the instability that arises from a change in the specification.

Design complexity metrics

Many studies point to the average number of bugs found in a design of a certain complexity. The state of the RTL and the number of bugs found are an indication of the complexity of the design. Many a time, a complex design is deemed to have more bugs than a simple design.

RTL Stability

RTL stability is a big indicator of how well the design is doing. This metric simply signifies the amount of time since the RTL was actually modified. If the RTL has been running a variety of tests and has passed the tests without modification for a long period, then the RTL is considered stable.

On the other hand, if the RTL has been undergoing changes and the last change was recent, then the question of the completeness of the RTL and verification arises. Most of the time, this leads to additional verification and validation being requested before the block is signed off.

Test case density

This metric indicates the number of test cases that actually address a functional object. While it does not imply that a small number of test cases are indicative of missed scenarios, however, if a test case undertakes to verify a large number of functional objects, *and the objects are not covered anywhere else*, then there is potential for a problem at hand. An example of this report is provided in the review section of the chapter *Putting it all together* in section 8.9.2.

Test Object density

This simple metric governs the number of test objects in a test. This metric is taken across the board and usually represents the quality of the verification environment.

Why test object density or the test case density an issue? it it even relevant? Test case density typically indicate how many tests actually exercised a particular functional object. It provides an indication of how well the functional object is tested. It is also a "free" metric that can very easily flag omissions or errors in a verification plan that has been executed.

On the other hand, test object density reveals the complexity of the tests. If the test object density is high, then there is a potential for risk from a couple of viewpoints. The first being, that if a certain functional object is invalidated or changed, then the test case must also be changed.

The second concern is that a reuse scenario may pose complications, if certain objects are not present. This implies that the test case must also be modified. Hence suitability of tests for reuse becomes an issue.

Test object density and test case density are two newer metrics that the author has been using in the past to gauge the state of the verification effort from a "blind" point of view, when

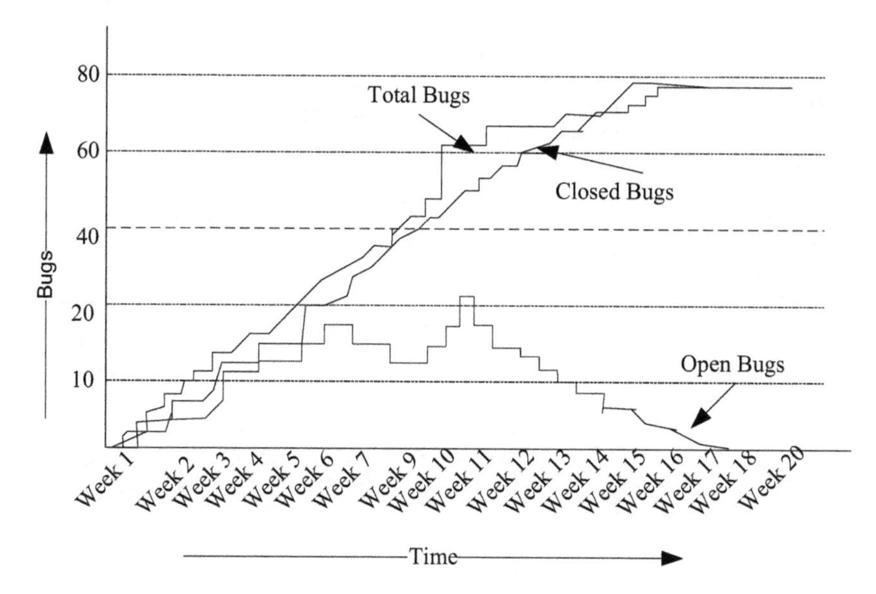

Figure 6.1. Bug curves typically observed in industry

the author had to conduct reviews in a rather short period with limited resources at his disposal. Invariably, more times than not, the author has found found tests that need to be written within a very short period by just observing the numbers available because of a correlation process.

Measuring functional object density and test case density is not very hard at all. If the concepts in automatic documentation and tagging approach described elsewhere in this book are followed, then the process of doing so becomes a fairly simple data-mining process.

Bug find rates

Bugs are being found in the RTL and in the environment at this rate. The typical Bug curves are shown in the figure 6.1.

When the bug find rate curve finally tapers off as seen in figure 6.1, it is usually an indication that the design is probably mature, as indicated in the diagram. A Tape-out is usually attempted after other parameters are satisfied.

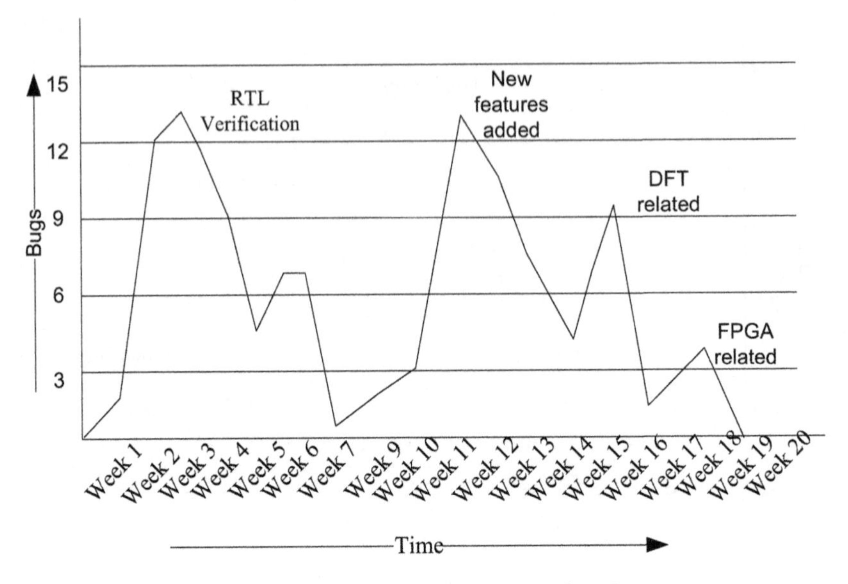

Figure 6.2. Delta Change in Bugs found

Bug saturation curves

In a similar vein, the bug saturation curves are an indication that the number of bugs found saturates after a period. As can be seen in figure 6.1, the number of bugs found in the beginning slowly tapers off as the design and the environment attain maturity.

A brief look at the figure 6.2 reveals that at the beginning of the project, the number of bugs found rises very quickly with over a short period. At some point in time, the number of bugs peaks, and then begins to drop. The drop in the bug find rate is attributed to the maturity of the RTL. In some cases, this figure may exhibit some interesting spikes in the curve as new functionality is brought online or a specification change happens.

Bug saturation curves are typically reviewed along with specification stability and other metrics in order to gauge whether the design is actually progressing towards its tape-out goals.

The rate at which bugs are closed is sometimes used as an indicator of productivity. In reality, What has been observed is that the number of bugs filed and the number of bugs closed over a period of time track typically well with a certain constant number of bugs open at any one point in time. This is indicated in figure 6.2.

6.4 Functional Coverage Metrics

With the advent of random generation techniques, it becomes imperative to determine the areas of the design that have been exercised with the test case. This brings about a need for a metric that describes how well the specific function in the design has been tested. This metric is called functional coverage. This coverage metric is measured by inserting specific coverage monitors throughout the entire design. Given the complexity of today's designs, the use of functional coverage metrics to reflect the health of a design has become commonplace and has gained much popularity recently.

Functional coverage needs to be instrumented across the entire design

One of the more important issues governing functional coverage is that the coverage needs to be instrumented across the entire design. For example, it is meaningless to discuss functional coverage if the coverage monitors are instrumented in only 10% of the design and the rest of the design has no functional coverage at all.

There are a variety of ways to implement functional coverage

Functional coverage can be measured by inserting assertions in the code, which can be used to log data to disk or update a scoreboard. The data can then be processed either during or after the simulation to reveal the functional coverage metrics being sought.

Many modern HVL's like e actually provide a cover struct and an analysis capability to analyze functional coverage.

Functional coverage is only as good as the points you define

While code coverage is measured regardless of the tests and the features being tested. In the case of functional coverage, various monitors have to be instrumented at the appropriate places. In addition, the monitor needs to be correct in order to ensure that the coverage is measured appropriately.

6.5 Structural Coverage Metrics

Code coverage comes in many forms. Code coverage can indeed be classified into many types.

Many of these coverage techniques are commonplace in the industry. In the earlier days, coverage was generally available by instrumenting the code with a special PLI/FLI that keeps track of the required parameter during simulation. Some of the modern simulators have this feature integrated into the simulator and offer this as an option during the simulation run.

Since code coverage is quite extensively covered in literature, the author has chosen not to include examples for each metric. A detailed collection of examples may be found in [1].

Line coverage

This metric usually describes the number of times that a particular line of code was executed.

State coverage

State machine coverage provides information on the visited transitions, arcs, or states in a finite state machine. This coverage can also provide the actual path taken through the state machine.

Trigger coverage

In some companies, this coverage is sometimes called event coverage. It indicates whether each signal in the sensitivity list has been executed or not.

Expression coverage

Expression coverage provides information about code that involving expressions present in the RTL. Expression coverage not only reveals if every line containing expressions was exercised, but it also tells you whether it was exercised in every way possible in the design.

For example, an "if" statement might be true in more than one way. Expression coverage indicates whether all the possible combinations of conditions were exercised.

Toggle coverage

This metric shows the number of bits that have actually toggled in the design, It is primarily designed for inputs and outputs and indicates if all the input and output bits have indeed toggled over the collection of simulations

Branch/path coverage

This coverage metric also indicates which case of the if/else branches were taken. Branch coverage, is an extension of expression coverage, shows which branches of an if statement or case statement were exercised. Path coverage is similar to branch coverage in that shows the exercised paths through a sequential if statement or case statement

6.5.1 Some Caveats on Structural Coverage

Does 100% code coverage imply anything?

Code coverage as a metric does not mean much. All it reveals is whether the verification code actually hit a portion of the design or not. When code coverage is run on a design, the results can typically be used to determine the portion of the design that is not currently exercised by the verification code.In many cases 100 %statement coverage may be completely misleading as evident from the example below.

In the figure A of Figure 6.3, one notices that If A = 1 and B = 1; then it becomes possible to get 100% line coverage. However, it only implies that we have 25% path coverage, However, one may get complete path and condition coverage using the matrix shown in the figure.

In a similar manner, If one uses a=0; b=0; c=0 and a=0; b=0; c= 1 combinations then we get 100% line coverage, and 50% branch coverage. One may add many more combinations to the same, and yet not achieve complete trigger coverage.

In case of toggle coverage, some tools report 0->X, X-0; X->1 and X->0 as well. One important thing to note for is that although toggle coverage reports a 100% coverage, there could still be a serious problem as illustrated in the figure.
Many more shortcoming of coverage approaches are discussed in [2]

6.6 Assertion Verification Metrics

Assertion Density

Using assertions in verification brings about newer metrics that are used to measure the quality of the ABV effort. The number of of assertions of each type in each module is termed as assertion density.

Figure A Line coverage Pitfalls

```
always @(A or B) begin
tempReg  = InitialVal;
if (A)
        tempReg = IntermediateVal;
if(B)
        tempReg = FinalVal;
end
```

A	B	
0	0	TempReg = InitialVal
0	1	tempReg = FinalVal
1	0	tempReg = IntermediateVal
1	1	tempReg = FinalVal;

Figure B Trigger Coverage Pitfalls

```
always@(a or b or c)
if ((a & b) | c)
WA = 1 b1;
else
WA = 1 b0;
```

You can get 100% line and toggle coverage. However, getting trigger coverage is not easy!

Figure 6.3. Line Coverage Pitfalls

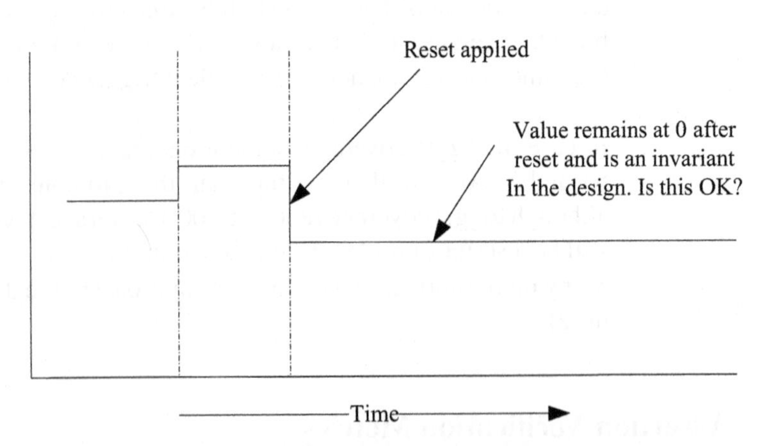
Reset applied
Value remains at 0 after reset and is an invariant In the design. Is this OK?
Time

Figure 6.4. Toggle Coverage Pitfalls

Proof Radius

Proof Radius Measures the amount of verification (in cycles) that was achieved by formal analysis. It is a measure of depth of the exhaustive search around each seed. The larger the proof radius, the more thorough the verification. A proof radius of 200 cycles implies that formal verification has exhaustively proven that no bugs can be triggered within 200 cycles of the initial state of the analysis.

DFV Seed rate

DFV can also be measured as the number of seeds as possible at shallow depth in a given time. The number of seeds per second per assertion at a small fixed proof radius is also a metric.

References and Additional reading

[1] Bergeron, Janick (2003). *Writing testbenches - functional verification of HDL models*. Kluwer Academic Publishers, Boston, 2nd ed edition.

[2] Jeffrey Barkley. *Why Statement Coverage Is Not Enough*. TransEDA Technical Backgrounder.

[3] The Linux operating system. *www.linux.org*. Also supplied by various vendors including Redhat.

PART III

MAKING VERIFICATION EASIER

This part focuses on helping the reader become more successful in verification. It is made up of many discrete sections. Each of the sections describes some care-abouts that help the verification engineer reduce the workload by attempting to structure verification workload in a optimal manner.

Some of the concepts presented herein are from various disciplines. All of the concepts have been time tested in a variety of organizations that build ASIC devices. Each of the concepts is treated in a discrete fashion. This will allow the reader to pick, choose or otherwise combine ideas to make the verification task go easier.

The author is confident that many of the techniques described herein have been successfully in use by verification engineers all across the world. Many organizations have these techniques as part of their "cookbooks" or processes. Some organizations use these techniques to drive their design and verification activity to ensure first pass silicon success with minimal difficulty.

There are no doubt a great many more techniques in use than what are described in this book. Some of these techniques are no doubt more significantly advanced than what is presented here. The author's hope is that this part will encourage others to share their experiences also so that we may all enrich one another.

There is a single chapter in this part titled *Cutting the ties that bind.*

Chapter 7

CUTTING THE TIES THAT BIND

Reducing Work in Verification

Looking at the global picture, the task of verification is indeed a much larger task than the design task. The design and development of a device involves translating the specifications down to RTL or lower levels. Verification on the other hand is all about making sure that the aspects that were not covered by the design and the proper operation of the design are covered as well. Hence, it can be said confidently that verification is indeed much bigger and more complex task.

Given that the verification task is huge, any tactic that reduces work is always welcome. This chapter explores the various methods by which the verification productivity can be improved. It offers some shortcuts that help reducing the verification burden.

One of the main inspirations for sharing the information in this chapter was to share some of these tips and tricks that the author has learned over a period of time. Some of these may seem intuitive in hindsight, but one of the things the author has learnt is that the *"The devil is always in the details"*and that small things do matter after all. Frequently, commonly known concepts are neglected and analysis and introspection in hindsight reveal that it had cost the project dearly in terms of effort or schedule or both.

It is noted that not all the concepts presented herein are applicable to every environment and there are probably many more good concepts implemented in practice.The author invites people to share these ideas in one of the more common forums on the internet.

This chapter is a collection of knowledge. There are many sections to this chapter, each different in theme from the other. The sections are developed to be independent of one another and may be implemented as desired. This allows the reader to pick and choose necessary optimizations that they choose to implement.

7.1 Considerations in a Verification Environment

Testbench is like RTL

Experience helps commonly made mistakes to be avoided! Experience also plays a role in formulating coding guidelines. These guidelines help insure uniformity in the code that is developed. Without these guidelines, many of the common mistakes usually wind up costing the verification teams time which sometimes translates to stress!

Nowadays, The designs are getting larger and larger. Many projects usually resort to using emulation or hardware accelerators to help achieve the throughput required for regressions on the project. Such an approach typically implies that a portion of the testbench actually resides in the emulator or accelerator. the author has actually noticed that the emulation or acceleration activity starts somewhere in the middle of the project, If care is taken early on in the development cycle, then the issues involved in porting the testbench to the emulator can be minimized.

Why is a synthesizable testbench even a concern?

One of the main reasons why some attention must be paid to this topic is that it is fairly easy to create a testbench that makes extensive use of non synthesizable constructs like events etc. When the partitioning of the testbench occurs, there will be extensive communication overhead between the simulator running on the host machine and the emulator. Under these circumstances, not much speedup will be observed. Initial planning and forethought could very easily avoid serious performance issues.

In one of the projects the author was on, the decision to even use an RTL accelerator was made late in the design cycle. When the testbench was ported on to the accelerator, little improvement was observed. When a code profile of a test run was analyzed, it became apparent that more than 70% of the time was spent on the host machine running some file I/O constructs in the testbench. When a good caching algorithm was used in the testbench, the speedup obtained was significant.

Take care before you move to an emulation or acceleration platform

The decision to move to an emulation/accelerator should be taken only after a profile of the code is taken to understand where most of the time is spent during the simulation. For instance, if 70% of the time is spent in some constructs in the testbench or in the simulation environment, and if a large part of the testbench is not synthesizable, then speeding up the remaining 30 % may not yield a great deal from an overall point of view. It is typically considered a good idea if coding standards for synthesis are also adopted for testbench and driver code portions that are synthesizable since the code can possibly reside in the accelerator/emulation platform if needed without any changes.

7.2 Tri-State Buses and Dealing with Them

Ensure that there is only one driver on a tristate bus

Tri-state busses are typically present in a verification environment when we have multiple drivers driving a bus. One of the drivers drives the bus and the rest of the drivers on the bus present high impedance to the bus. By far and large, it is preferred to have a single interface from the testbench side to deal with the tristate bus. This typically helps avoid bus contention. In some circumstances, this may not be easily possible. Ensuring using some mechanism that the drivers are unique or one-hot simplifies the debugging effort.

Add an internal signal to the driver to let you know the driver is driving the bus

A long time ago, the author has had the (mis) fortune of configuring a couple of drivers on the bus without any appropriate debug signals that let the author know that the bus was being driven! The only recourse the author had at the time was to run the simulation to that point where a failure was suspected and then use a step-through on the simulator interface to reveal the drivers at every instant on the net that was being driven by multiple drivers. This was a time consuming task and was difficult to manage since the failure occurred after a fair amount of time was lost in simulation. In hindsight, the author believes it would have been far easier to insert a signal and debug it by observing signals instead[1].

7.3 Dealing with Internal Signals

Internal signals are one of the "necessary evils" in a verification environment. Sometimes, the I/O pins of the device may not be able to provide sufficient information to allow the testbench to perform certain tasks. Under these circumstances, it may be needed to access some internal signals to simplify the task of developing the testbench.

Ensure that all internal signals are maintained in a single file or location in the environment.

One of the main considerations that need to be made when a test references an internal signal is whether the test needs to be portable or reused at other levels. Such considerations usually make the choice a little easier to make. Using such an approach enhances maintainability. Further, if for some reason, an extra layer of hierarchy is added to the verification environment, it becomes a very daunting task to go through several files to actually add the correct hierarchy to the environment.

Ensure that synthesis does not lose the signals that you created.

Another of the bigger concerns about referencing an internal signal is that signals are usually wires which aren't usually preserved during the synthesis operation. The synthesis tool may very well discard the very nets that are being observed. This brings about the problem that the signals need to be maintained a great deal. Such maintenance is usually non trivial.It is noted

[1] There are probably other simpler ways to do so, but at that time, other ideas hadn't occurred to the author.

that module boundaries may be used in some circumstances if the synthesis tool has been specifically configured to leave the module boundaries alone.

Typically, the design goes through a stage of development followed by a stage of maturity. During the development phase, the RTL may experience significant volatility depending on various circumstances that are present during the development of the RTL. If a test were to refer to some internal signal then the test will have to be modified every time the signal is changed or moved.

Don't force internal signals unless you absolutely need to

Doing so can be quite dangerous since forcing the signal can indeed mask some failure that would otherwise not have been detected. This practice has its share of pitfalls. There have been occasions where the regressions were declared as passed till a final review was done and then the results were declared to be invalid when it was discovered that some of the signals were forced. The result was that the tests had to be rerun and re-certified before the tape-out of the device.

Such an approach can be costly if it is not caught early. On the other hand, it may not be possible to create a specific set of scenarios at a certain level of integration if internal signals are not forced to some value.

Keep all the list of internally forced signals in a single place

The forced signal list needs to be maintained and periodically reviewed to ensure that only the absolute minimum signals are indeed forced. Keeping the list in a single place enhances maintainability. It also speeds up the review process.

It must be possible to account for each signal in the force list with an explanation as to why the signal cannot be removed from the force list.

In some circumstances, there will be various lists of signals that are forced in different environments – RTL/Gate being one example. Different block level environments also may have lists of signals that are forced. Sometimes there are too many of them to deal with.

It is recommended that at the very minimum, a list of such signal lists be preserved. This will ensure that all the signals in the list are indeed tracked to ensure that there are no false positives generated by the forcing of the signals.

7.4 Environmental Considerations

Keep the environment easy to use.

One of the key aspects of a verification environment is that it must be easy to use. Simplicity breeds reliability in the environment.

Design engineers will use the environment as well as verification people. It therefore becomes apparent that the people who wrote the tests may be different from the people who wrote the environment.

What is the consequence of the verification engineer not following this guideline? Simply put, the design engineer has enough to worry about. If the environment is large and complicated, then every time, the designer will turn to the verification engineer for help in running the tests in the environment. As a result, the verification engineer is now burdened with not only maintaining a complex environment, but also effectively being at the beck and call of the designer to help the designer run tests!

There is sometimes a tendency to use non standard debug tools or use macro's extensively. Doing so makes debug of the environment very difficult and complicated. Usually, this works against the engineer who has developed the verification environment – *No one can help the engineer who builds such an environment since it is his own creation.* The author has been in a situation where every small modification to a set of files had to go through a specific engineer simply because of the complex macros the engineer had embedded in the environment. It was impossible to even add a couple of lines to the configuration file so that one could run a couple of tests to verify a few ideas before asking anyone for help! Needless to say, there was considerable confusion and a stifling of innovation.

Simplicity and ease of use also have other important implications. Engineers can be brought on board faster. In the beginning of a project, a few engineers typically wind up developing major portions of the environment. They then involve other engineers in test-writing and debug activities.

Make sure the environment is debug-able using the same tools that areused for RTL development

The environment is a combination of RTL and the testbench. The environment is developed to test the device. Hence, it must be concluded that there is the possibility that there are bugs in the environment as well. It must be possible for anyone to quickly identify a bug as being present either in the RTL or in the verification environment. The most difficult thing that can happen is that the RTL designer states that there is no bug and the test fails and all the pressure is now on the poor verification engineer who has to debug not only his environment and tests but the RTL as well. Being the only one who can perform the task makes it even more difficult!

Capture all relevant data

One of the basic features that must be present in the environment is that it must be able to capture all the relevant data that would be required to debug the test case if failure occurs Capturing of relevant data must include at the very minimum the following:

- The seed and other data required to reproduce the run
- Any command line options that were used
- Any generics that were used in the test run
- Any Output from the simulation
- Output from debug and logging monitors

When a test case fails typically, the debug process usually begins with parsing of a log file to see if any relevant clues that would indicate point of failure are present. Sometimes, it becomes impossible and difficult to log all the appropriate data into the log files because of the fact that the simulations are long and the log files that are generated are huge. In such circumstances, the author recommends that extensive use of debug levels can make the problem a lot more manageable.

Make sure that the environment supports both stand alone testing of the RTL as well as the environment

The environment is like RTL. If you look at it another way, the testbench is effectively a piece of code that is basically used to test another piece of code. That is not to say that the testbench is completely error free. Most engineers realize that the test-bench and the environment itself have its own share of bugs. Hence, it will become necessary to test either the environment or the RTL using some specific directed test cases. These test cases may choose to excite specific features of the environment as well as the testbench in a particular fashion.

Ensure that you dump only the minimum data into the waveform dump file

Under some circumstances, it might be necessary to rerun the test case with some additional waveform dump to a file so that the actual problem can be debugged properly. It is apparent that a large waveform dump file will definitely take a little longer to be created. In every project, disk space is always at a premium. Often, the waveform data is kept around for a long time for certain reasons. The practice of saving some golden waveforms from a simulation which had considerable time spent on it has been observed frequently in many organizations. The duration of the waveform storage has indeed varied though!

Being able to create a waveform with the right data and the smallest size is nowadays possible with various tools that are now available to manipulate the waveform dump files. This feature must be used very effectively to prevent issues in Disk management.

If the reader does need to keep a wave dump file around, it is recommended that it be trimmed to the smallest possible size. The author recommends saving a complete snapshot of the environment instead. This snapshot could be a version label that reproduces the wave log file in question.

Providing options to the user to be able to select exactly what is dumped into the waveform dump file is critical. Having too little data dumped with a certain option into the file leads to the consequence that the user chooses the default – Dump everything and worry about it later. This compounds the problem of already hard to obtain disk space.

Partition test code based on functional objects

At the very minimum, it must be possible to partition test code and associate the each piece code with a certain functional requirement. These functional requirements are termed Functional objects [2]

Why is such partitioning important? An effective partitioning enables the possibility of identifying tests and environment code that is part of the verification environment. If features are modified or modes are trimmed from the device, it becomes a straightforward matter to identify the affected portions.

Time pressures are always a factor in any verification effort. Many a time, the schedule for tape-out takes precedence over the completion of verification. This is a fairly common occurrence. When this occurs, one can prioritize the tests that need to be developed and maintained over other tests.

It now becomes possible to review test code and associate that code with a functional requirement that is being tested.

In addition, the identification and preservation of tests that give the most contribution to functional coverage is also enabled. (test grading practices).

Make all tests be self-checking

There are several ways to test that a test indeed did create the scenarios that it was supposed to and that the response was appropriate. The most common methods are:

- *Eyeballing:* Looking at the test results visually to determine if the test did indeed accomplish its goals.
- *Comparison with some expected output:* In this method the test output is compared with some expected result which has been certified as correct.
- Using a program to determine the correctness or failure of a test.

Experience has often revealed that the introduction of the human element brings about opportunity for mistakes to be made.

[2] Functional objects are described in detail in the chapter putting it all together.

A computerized decision on the other hand does not suffer from this deficiency. Consequently, It appears apparent that any test must be created in such a manner that it allows the test to be checked automatically. If it becomes possible to determine the results of the test from some data, then we can make this test regressable.

Automate the checking mechanism as far as possible

Another key environmental consideration is that the mechanism for checking the tests must not have any manual intervention from the test writer as far as possible. This approach eliminates the possibility of human error. In a typical verification project, the same test may be run many times. If the test relies on the human element to ensure the pass or fail of the tests, the possibility of errors rapidly increases with the number of tests and the number of runs of the same tests. This essentially limits the number of tests and scenarios that can be tested since the human element can only process a certain number of tests in a given amount of time.

Sometimes it may not be possible to automate all the portions of the checking mechanism. For example, when a mixed mode simulation is being run, it may become necessary to look at the spectral analysis of the circuit output. Such tests are hard to handle in any project. In such situations, it may be possible to check some portion of the output with an automated mechanism and limit the amount of manual checking that is done.

Make sure that all simulator warnings and errors are identified and addressed

Simulator warnings are usually something that needs immediate attention. The warnings from many commercial simulators can be turned on/off. However, a detailed observation of the warnings can often provide clues to the failure. This is one of the first places to look at while debugging a simulation. (Old saying: *When all else fails, read the manual! In this case the log file should never be the last one to be looked at!*).

Warnings also serve a second purpose. They reveal to the user the problems that the simulator found in the code. Usually, simple issues like mismatching bus widths or redefinition of some parameters are printed as warning messages.

It is important to ensure that all simulator warnings and messages are accounted for. Doing so ensures that the most obvious errors leading to unusual/unexpected behavior are caught early on in the game.

Create a testbench so that both RTL and gate simulation are supported properly

In most projects, the RTL is typically the first to be developed and debugged. Once the debug is complete, the RTL is typically synthesized. The Gate netlist that is built is then used for simulation runs. Usually, the Gate netlist needs a little more work than the RTL netlist to operate correctly. Some things that worked in the RTL netlist may no longer work in the Gate environment. A notable example is the forcing of certain signals in the design. Depending on the synthesis approach, the hierarchical nature of the RTL netlist may or may not be present in the Gate netlist. Another interesting thing that the reader may find is that sometimes the RTL design is done in VHDL and the GATE netlist possibly is in Verilog. If this is not accounted for, one can be assured that complications will arise if some VHDL specific aspects are used to construct the environment.

Under these considerations, it is recommended that the testbench be constructed in a fashion to allow the RTL and gate netlist to be used from the same testbench. This approach will save the effort of developing a separate testbench for the gate netlist and make sure that the testbench components that are used for the netlist are in sync between the RTL and the gate netlist.

Ensure that a subset of all RTL stimulus will operate on gate level models

One project the author had an opportunity to observe had the entire design in VHDL. The simulations and the environment were all designed to operate using VHDL. However, when the gate netlist arrived, the netlist was completely in Verilog. This caused some difficulties when the gate netlist was to be used for simulations. Some of the portions had to be completely rewritten to accommodate this change. If this fact were known and accounted for in the beginning, then the entire process of modifications and iterations could have easily been done away with.

As the reader will observe, not all the tests from the RTL environment will be run on the gate netlist since runtime is usually a issue. Selectively picking the right set of tests will help the verification engineer to ensure that the design is adequately verified.

Testbench should drive the I/O connections of the module or the device

One of the most important environmental considerations is that there should be no signals from the testbench driven into the internals of the device under test. This guideline is similar to the guideline for internal signals. Doing so renders that some portions of the logic may be untested or even lead to false results. The author's opinion is that all statements on the quality of the design are not valid if the testbench is driving and reading internal signals.

In the real world, It is not possible to connect to the signals inside the device. During the initial stages of the development, a few internal signals are used to speed up testbench development when the RTL is not mature. These signals must be removed as soon as possible.

During a review, this is one of the most important things to look for when ensuring that the device is indeed tested correctly.

7.5 Dealing with Register Programming

In the hardware world, All information that is communicated to the outside world is done either through pins at the periphery of the device or via a set of registers on the device. In a typical device under test, the registers are programmed to set up some specific parameters of the device.

In this section, we explore the ways that registers are programmed and also offer the reader some possibilities to simplify the register programming effort.

Register programming has a large impact on the testing of the device. A large portion of the verification engineers job is usually dedicated to having the right settings for the device. In his career, the author has noticed that many times that the verification engineer had to rework the tests time and again since the engineer did not choose to use a programming style that would support any changes in the register specification.

There have been many instances in the authors career where the register specifications for the bits have changed. In some instances, the register was moved to better accommodate the

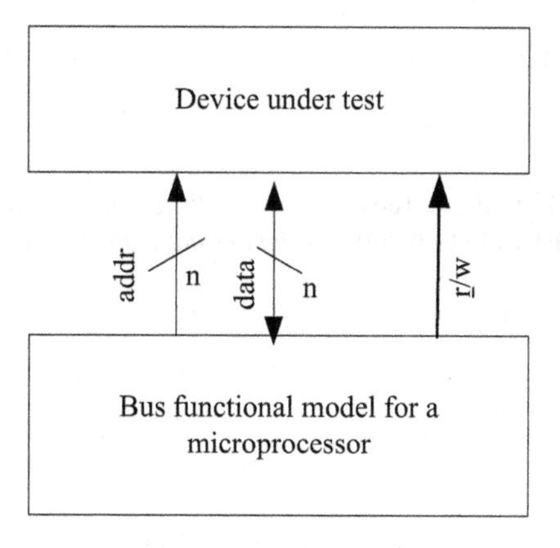

Figure 7.1. Register Read Write

designers needs. In some cases, the register bits moved positions. In others, additional bits were added. In many of the cases, the tests needed updating to reflect the latest register specification.

Given all this change, it becomes imperative that the tests do not incur additional maintenance costs due to flux in the specification.

For example: Consider the device in the figure 7.1. This device is shown to have a microprocessor bus with a bidirectional data bus and read/write signals. The device in this example is considered to have about 500 bit settings. This is a large number, though not uncommon with telecommunications or other devices today.

In the example in figure 7.1, we assume that the device has a memory map from 0x0000 to 0xfff0. The figure also shows a bus functional model of the microprocessor. This bus functional model interfaces with the hardware of the device under test.

If one were using transaction based verification typically, the bus functional model offers a couple of transaction interfaces to the microprocessor interface.

```
task write_all_registers;
begin
    -----
    ......
        do_write_UP1(0x0001,1001); // write to the outgoing and incoming enable bits
        do_read_UP1(0x1119,0000); // read the status for the device.
    ---
    // additional test code.
    ....
    ...
end
```

Figure 7.2. Read and Write

a typical example of the write and read transaction in a test could be shown as:

An analysis of using this coding style indicates the following:

It is a simple style to write tests in this fashion. However, tests using this method are not portable or easily reusable. If a bit is moved or the register moves during some course of the project, then the test will need to be rewritten.

If the device or a portion of the device is re-used in another derivative device, then in all likelihood, the test will be thrown away. At the end of the project, the only way to find out all the settings that have been exercised by all the tests is to go through and audit the tests or rely on some other metrics.

Other alternatives

Some of the verification environments on the other hand use an indirect scheme. The register name is built using a mnemonic list that is then used to program the device. This approach solves a few of the concerns of the previous approach.

This approach is still commonly used by many environments. However, it still is impossible to identify from the tests the bits that were set/unset without a copy of a bit map next to the engineer! The process of referring to the bitmap is usually time consuming.

```
task write_all_registers;
begin
  -----
  ......
     do_write_UP1(ENABLE_REGISTER,0x1001); // write to the outgoing and
     incoming enable bits

     do_read_UP1(STATUS_COMP_REGISTER,0000); // read the status for the
     device.
  ---
  // additional test code.
  ....
  ...
end
```

Figure 7.3. Alternate read and write example

If this type of register programming is in a few tests, the amount of work will explode! The approach does not scale well in case of certain types of devices (for example networking devices where there are thousands of bit settings). The approach is also not portable between device versions. In addition, it is still impossible to determine statistically how many of the register bits have indeed been exercised during the course of the verification effort.

There are variants to the above scheme that are widely used in verification environments today. In the critical analysis above, no statements on the validity of the approach is inferred. In case of both the methods in question, the designer and the verification engineer need to do some work in order to cope with the design changes. In the initial stages of verification, this change can be quite a bit. It can also be frustrating at times.

7.5.1 A Hybrid Approach to Register Programming

Modern languages like *e* or *Vera* now provide the ability to create data structures for registers. if we consider a register and a bit as a class object, we observe that the bit field itself has the following properties:

- Value on reset.
- Default value.
- Offset from zero.
- Size (number of bits) etc.
- Read/only or similar attribute.

similarly, amongst other properties, the register itself will have the following properties:

- Value on reset.
- Default value for undefined bits.
- Offset from register base address.
- Address.

It now becomes completely possible to create a collection of simple data structures that address each register and bit by its name as shown below.

A register can then be constructed using a structure of bits and some additional properties as indicated in figure 7.5.

Using the coding style in the figure 7.6, one can now create register programming sequences that look like that indicated in 7.6.

There are many good examples and discussion of this code in [9] and [10]. The approach indicated is not the only way to address the register programming problem. There are many other ways to achieve the same result. Some commercial tools like [11] or Cadence register package [12] offer some additional features that may be deployed effectively by the verification user.

There are many advantages to using a scheme as above. These are described below.

```
// The definition of a bit is below
class RegBit
{
public:
        RegBit(void);
        virtual ~RegBit(void);
        void SetBitValue(unsigned short BitVal);
        void CompareValue(usigned short value);
        // initialization when the class is created.
        void Init(unsigned short BitVal, unsigned short BitMask, unsigned short
BitShift); //void SetRegPtr(class Register *regPtr);
/* various elements to signify
value on reset
default value
Offset from zero
size (number of bits) etc.
read/only or similar attributes
*/
};
// The definition of a register is below.
```

Figure 7.4. Bit Example

A static analysis is now possible

If the reader were to observe, the power of a name now becomes incredibly evident, If the engineer were to factor out the basic register read and write tests which have all the combinations, the ones that have been exercised and tested in the tests are now completely self evident. A simple text search and sorting algorithm can easily identify issues in the verification.

If the register bit settings were to control certain functionality, then the extent to which the functionality is coded is evident at a glance!

Debug becomes a simpler affair

Debug is one area of verification where time is spent in copious quantities. One finds often that a test fails because something was not set somewhere.

The designer frequently asks the verification engineer for help in debugging the design since the designer needs to look at an address map or a bit map repeatedly for a debug. Using a name convention ensures the “reduce, reuse, recycle” habit.Once the

```
class MyRegister{public:
        // Class Construction
        MyRegister(void);
        MyRegister(unsigned short Offset, unsigned short Value);
        // Class Destruction
        virtual ~MyRegister(void);
        //Operators
        // Class Initialization
        void Init(unsigned short Offset, unsigned short defaultValue);

        unsigned GetOffset(void);
        unsigned short GetValue(void);

        // Set Methods
        void SetOffset(unsigned short Val);
        void SetValue(void);
        void SetValue(unsigned short Val);

        // General Methods
        void PrintReg(int opcode);
        void CompareRegister(unsigned short mask, unsigned short shift);
        void CompareRegister(int Value);        // Instances of bits in this register

// Instances of Bits
        RegBit bit1;
        RegBit bit2;
        RegBit bit3;
        RegBit bit4;

// Various initialization sequences are included and not shown

// Various housekeeping elements
// value on reset
// default value for undefined bits
// Offset from register base address
// Address
};
```

Figure 7.5. Hybrid Register example

particular register is debugged, it is clean for ALL tests in that design. One does not have to visit it again.

```
Register1.bit1.SetRegister(0x1);
Register1.bit3.SetRegister(0x0);
Register1.bit2.SetBit(0x1);
....
Register1.SetAllValues(0x0010);.
```

Figure 7.6. Hybrid Register Programming Example

The naming convention also ensures that a glance through the test files usually reveals if some mis-programming of the device happened!

Relocation of registers and bits is now possible

One of the nice things about using a naming approach that uses a object mechanism as shown in the preceding discussion is that it now becomes to make the test completely insensitive to bit movements in the design. For the most part, there is not much movement in the bit locations once they are designed.

However, consider a scenario where an IP that has been verified without this approach used for a certain version of a device that has been successfully taped out. Let us consider a derivative design that needs to modify the IP so that some additional controls needs to be added and as a result some control bits need to move around.Its certainly possible that the address map would possibly change as well.

If care wasn't taken in the beginning, it would be reasonable to assume that most of the tests exercising this IP were going to be affected. Interestingly the approach ensures that a straight recompile of the code with an updated header would probably take care of the issue.

7.6 Dealing with Clocks

Create clock modules with parameterized delays for control and skew

Use a single constant or a variable, that is in the clock module that can be used to scale the clock up or down. This makes the changing the frequency or the design or adapting it to a different design much easier.

Be careful of rounding and truncation in clock expressions

Use of division instead of multiplication can cause rounding errors based on timescale and resolution. For example: Consider the use of a simple clock at 622 MHz. This frequency is typically used in some telecommunications devices. If there is the need to generate the following frequencies: 311 MHz, 155.55 MHz, 77.75 MHz, and 4.8 MHz, one of the common things to do is to define the highest clock at 622 MHz and then use a set of successive dividers to arrive at the lowest frequency.

Interestingly, this causes errors as the frequency is lowered and successive division takes place. This is particularly true if timescale and precision are not chosen appropriately.Hence it is recommended to use multiplication using the appropriate multipliers. A good example of this is given in [5].

Treat derived clocks with care

A derived clock is generated by a flip-flop, latch, or any internal clock generator (such as PLL's, frequency dividers, etc.) in the circuit. Derived clocks can render a large part of the design un-testable later on, because the flip-flops are not controllable from any primary input. Derived clocks usually lead to DFT headaches as well. Another important issue to watch out for is that the derived clocks must be generated in the same process block (always in Verilog) to take care of skews that may be unintentionally created.

Model the skew on the clocks in the clock generator module

Many a time, clock skew modeling is typically done later in the verification cycle when the design is stable. Typically, however, the clock skew, jitter and other parameters are usually available early on in the design cycle when the product specification is available.

The author recommends that the appropriate skew and other parameters on the clock be modeled early on and instantiated in the testbench. For the most part, these can be turned off during the initial testing. The parameters can then be used for verification as and when deemed necessary. The approach also helps the testbench to avoid changes at a later date when the design and the testbench are stable after having passed many regressions.

Do not drive any clock inputs deep in the design

Driving a clock input deep into the design is akin to using a pseudo internal signal. The design must be able to generate any clocks needed to drive the internal nodes of design as a relationship of signals that are present on the external pins. If a clock is needed deep down inside the design, the author recommends that the reader file a bug on the design!

7.7 Driving the Design

Each driver has a single function. - Drive one interface

For the most part, a driver must completely cover the interface it is designed to be with. If commands for the particular interface are not generated, it is important that the driver flag the command as an error. Otherwise, the user of the driver will spend a lot of time in debug which may point back to the driver and result in loss of trust in the driver. One thing to avoid in a complex protocol is to keep a subset of driver commands from one driver and another subset from another driver. A corollary of this is that it is important not to have different transactions coming from different drivers for the same interface merely because it is easy to implement in the initial stages. This also implies that the driver must be designed properly. An example of how to do so is derived in the next chapter.

Keep the driver modular and simple

The other key requirement is that the driver in itself should not have any instances of another driver in it. This makes the driver hard to port between environments.

Drivers only drive boundary signals and not internal signals

Drivers or bus functional models must only drive the primary external signals. For the most part, the driver will not be usable properly if it is referencing internal signals deep in the design and using some particular protocol that is specific to that interface. If the design moves from an RTL to a gate netlist, then it is possible that the driver is unusable.

Drive Inputs only as long as you need to

Another key aspect of the driver is that the driver should only drive the nets or ports of the design only as long as they need to. They must not function as "bus keepers" by holding the value longer than it is necessary. It is highly possible that such behavior will mask an accidental bug.

Make it possible to disable drivers from a single location

It is important to provide a mechanism to disable the driver and this disable mechanism must be centrally located. It can be called by all the sources that need to use this driver. One may choose to use some global variables which are then mapped to each driver specifically to control the drivers.

Do not mix driver and checker functions in same file

This guideline comes from a mistake that the author once made. The code for checking a protocol relied on the same signals that the driver was using. In order to save time, the author coded both into the same file little realizing that the practice would create problems later! The author then had to port the checker over to the next hierarchical level in the design and had to rework the checker to fix the mess!

7.8 Debugging the Design

Debug always takes a significant amount of time

One of the main factors in a debug effort is that it takes quite a bit of time. The paragraphs that follow describe general debug advice that is gained by experience gained by many over the years in verification. The author is sure that with the advent of modern tools and techniques which are many, there is much that can be added to this list to make it more comprehensive as a guide to debugging designs.

Note that a wave dump file must always be the absolute last resort when debugging a design. Unfortunately, it is sometimes the first thing that gets looked at!

It is important to find the root of the problem and get rid of it

Almost all verification engineers have heard this often repeated mantra "get to the bottom of the matter". However, doing so is quite challenging. The biggest challenge is to identify with certainty the actual cause of the problem and deal with it properly.

The author has observed a bug filed in a bug tracking system making the "bug rounds". The bug is typically filed in response to a test failure and deemed to be a certain problem. It is then recategorized as a different issue and then another. Pretty soon the bug finds a life of its own! Jumping from owner to owner.

At the end of it all, it becomes pretty much impossible to search for the bug in a system if it was not categorized properly in the first place. In the light of the , the author recommends that the original bug be closed with a new bug entered in the right place.

Places where the problems could be

The most important thing before beginning a debug is to identify what had changed from the previous time the test ran. If a test is written and run against the RTL for the first time, the problem could be anywhere. This poses a significant challenge. Many people tend to triage issues by using some sort of an algorithm appropriate to the situation.

The most important item is to understand the scenario that is being tested. After this is done, the triage is done to eliminate issues with the following broad categories.

- What changed since the test ran last time around?
- Test code that was developed.
- Any similar outstanding bug.
- The RTL itself if the code was untested.
- The environment causing issues.
- Non repeatable issues.

And so on. The idea being to quickly determine if some known problem already exists or if a new problem has arisen. The common method of identifying problems is to typically start by looking at the log files for obvious signs that something is wrong. Any well written environment will have detailed logging controlled by verbosity levels. Assertions that have been embedded in the design that fail help identify the problem quickly.

An encoding scheme in the log file which reveals the details of where the transaction is coming from is usually helpful.

Many tools today have powerful active annotation and debug features that allow the user to identify the source of the problem. X propogation problems or incorrect values are quickly identified in this approach.
Debug levels have been discussed elsewhere in this chapter. At the very minimum, an error may or may not stop the simulation using a flag controlled by the user depending on the particular situation.

Use a coding style that tells you where the errors are coming from

The author would like to suggest that any error flagged by either the testbench or the design use some sort of a coding style that uses an error number or something similar so that a unique string can be attached to it. This sort of style is very prevalent in the software industry.

For example: a range of numbers can be given to each block in the design. When the messages are printed in the log file, a script can help to parse the messages for each block and help debug.

Lockup or Runaway detection is a must

There are many scenarios where the test case or testbench may be hung up waiting for an event to happen. In order to determine if this is indeed a problem, a watchdog timer that terminates the testbench is useful to help the testbench to terminate. An error message must be put out dictating why the simulation was terminated and what the testbench was waiting for. This approach helps the debug effort.

Do not use keywords like error/warning in the stimulus

Most parsers and filtering mechanisms wind up using error or warning as standard keywords.These filters will then either trigger false results and this will cause unneeded confusion.

This is a simple guideline. The interesting thing is that a keyword like "error" or warning is something intuitive. The interesting thing is everybody is yet to come across a project where this guideline is not violated!. Someone or another usually winds up breaking this guideline and wreaking havoc on the regression results at least once during the lifetime of the project!

7.8.1 Making Note of Errors

Use common routines to report errors

A key guideline in error reporting is that a uniform error format typically helps to simplify scripts that parse the log file and also helps speed up debug.

There is usually a script that parses the log files and determines the pass/fail result of a test. Having a uniform format simplifies the parser script and makes it easy to implement.

Errors must be detected at the point of failure

One of the important aspects of detection of errors is that the error must be detected within a few cycles of the point of failure. The testbench should also allow the simulation to continue based on a user configurable switch.

Failure detection could be by many methods. Using assertions in the design is one of them. The test bench should be able to detect and terminate the simulation even if several monitors are instantiated and also be able to provide the exact location of the fault. This aids in debugging.

7.8.2 Debug Levels

There is extensive mention of debug levels at many points in this book. The concept of using debug levels is a fairly common concept that has been borrowed from the software world.Used wisely, it can encapsulate debug messages at different points to help identify the possible source of the problem.

An example of a debug level with some pseudo-code is shown below. There are three debug levels in this example. As can be seen from the example, one can control the amount of output by choosing DEBUG_LEVEL_1, DEBUG_LEVEL_2 and DEBUG_LEVEL_3

```
ifdef DEBUG
        ifdef DEBUG_LEVEL_1
                display the monitor level
                display some packet statistics
        endif
        ifdef DEBUG_LEVEL_2
                More display code
                display some monitor output on level and time
                display some more monitor output and number of
                packets so far.

                log all transactions into  a log file with full details of
                the start and end of each transaction

        endif
        ifdef DEBUG_LEVEL_3
                display some monitor output
                display some more monitor output
                log all transactions into  a log file with full details of
                the start and end of each transaction
                print all the packets into the log file.
        endif

endif
```

Figure 7.7. Debug Levels

Plan out the debug level implementation

Debug levels can be implemented easily when the verification code is being developed. Alternatively, it can also be added after the core pieces of the verification environment have indeed stabilized. There are usually different contributors to the verification environment, what is important is that the definition of the debug level is uniform across all the implementations of the verification environment.It is important to have consistency in the debug environment implementation. It is crucial that a certain debug level has the ***same*** definition regardless of its location in the project.One can very well imagine what would happen if DEBUG_LEVEL_1 in the example meant verbose in one module and no verbosity in the next, and medium output in a third module!

Always have one level where all debug is turned off

Debug is intended to increase the verbosity of the output. It is important that there be a single level where the regressions can run with minimum verbosity.

For the most part, there is a balance that needs to be struck between two conflicting requirements. The requirements are between being able to debug a failure in a regression which implies maximum data is logged, and running as fast and efficiently which implies as little data as possible is logged. This is particularly true in the final stages of regression, most of the time, all that matters are the results of the test in the regression.

As run times increase to be in the order of days, the log files can get huge and sometimes exceed file system limits. Having to rerun the test again with a debug turned on is another challenge altogether. Use of controlled verbosity levels as illustrated help the user to identify the problem quickly and then address it quickly.

7.9 Code Profiling to Keep the Inefficiency Out

Many simulators today offer some advanced tools that help the verification engineer identify where the simulator is spending most of its time. This is called profiling. The code profile reveals the area or segment of the code where the simulator is spending most of its time in relation to the overall simulation time.

Such a report is usually available automatically from many tools. Many simulators can provide this output with the mere application of an option.HVL's like *e* also prove a tool that allows the user to find out where most of the time is spent and how frequently the particular piece of code is accessed.

Make sure the testbench isn't in the profile!

The example in figure 7.8 looks like a fairly simple piece of code written in Verilog. The task reads a file and places the data in a variable called packet_data. It uses a PLI call to get the data from the file.

Code profiling is a powerful tool. Code written in a certain way may inadvertently cause the simulator to spend quite a bit of

```
task get_bytes;
  input [31:0] fileptr;
  reg [31:0] char_read;
  reg [7:0]  data;

begin
  packet_count = 1;

  char_read = $fgetc(fileptr);
  while (char_read!=32'hFFFF_FFFF) begin
    //1st Byte
    while (char_read[7:0]==8'h0A ||                          // NEW LINE
              char_read[7:0]==8'h5F ||                        // UNDER SCORE
              char_read[7:0]==8'h20)                          // SPACE
      char_read = $fgetc(fileptr);

    if (char_read!=32'hFFFF_FFFF) begin
      // Checking for EOF after the SPACE or NEWLINES.
      data[7:4] = convert_hex(char_read);
      char_read = $fgetc(fileptr);
      data[3:0] = convert_hex(char_read);

      packet1_stream_data = data;

      // Reading the Next Character.
      char_read = $fgetc(fileptr);
    end
    else begin // ON EOF exiting out of task.
      disable get_bytes;
    end

    @(posedge packet1_clk);
    #1;

  end
  packet1_stream_data= 8'h00;
end
endtask
```

Figure 7.8. Code Profiling Example

time, something that the verification engineer did not actually intend. On observation of the code in the example, one observes that this code is functional and does not pose a problem if it were studied in isolation.

Now if one were to consider using this task to read data from say eight files to feed data to a device with eight different ports, one notices that the simulation becomes very slow because of the file I/O that is part of the testbench.

In any simulation, the most important thing to ensure is that the simulator is spending as much time as possible sending cycles or events though the design. Any time spent in the testbench is in effect not utilized in sending cycles through the design. The testbench may be preparing the data to be sent through the design, But the clock has not stopped ticking while the testbench is doing so.Hence the total time for the simulation is the sum of the time spent in simulating the design and the testbench.

In order to maximize the performance of the environment in simulation, it is recommended to make efforts to ensure that the simulator spends most of its time simulating the design rather than in the testbench. If one were to use the code from the above example in an hardware accelerator, then one may see very little speedup offered by the accelerator compared to conventional simulations.

Be careful when you interpret the numbers

One has to read the profile numbers carefully.The quantity of the testbench code in relation to the amount of device code must be taken into account when reading the profile data. If the device under test is very small, then the testbench code may dominate the profile. Then optimizing the RTL may be quite the wrong thing to do!

This sort of scenario is very common in module level environments where there are a large number of checkers and monitors and the device under test is very small. It is recommended to review the profile more in subsystem and system level simulations where the run time of simulations is usually a concern.

7.10 Regression Management

Report a regression summary

Management of regressions is typically assigned to one member of a verification team. This member typically runs periodic

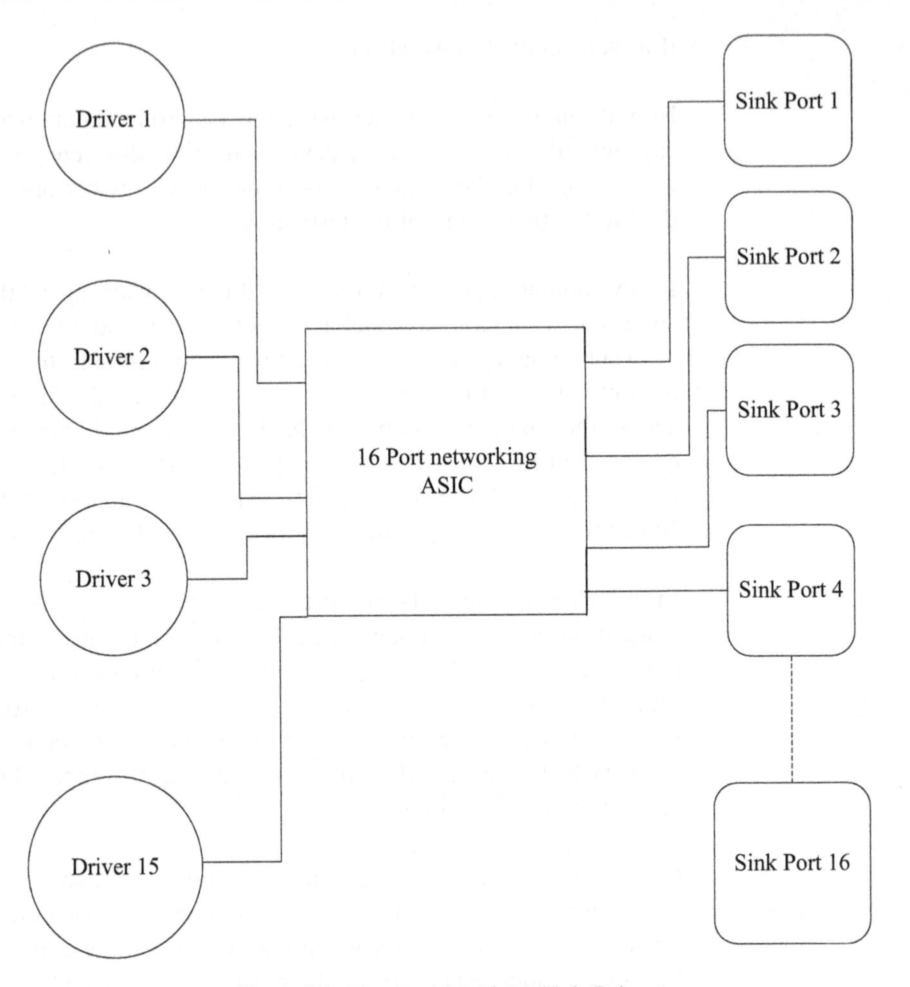

Figure 7.9. Test Bench with Multiple Drivers

regressions and reports the results. Sometimes, it is a rotating duty handed by all team members. Many a time, a special user is created on the network to handle the regressions.

The tests in the regression finish at different times. Since many members would like the status quickly without waiting for the results, It is suggested that a periodic process or program be used to automatically generate the state of the regression. This can be accomplished simply by using a cron [3] job.

[3] Cron is a facility available on UNIX systems. It allows the user to run some programs periodically.

Such a practice has some advantages. If a member of the verification team were to notice a problem, then action can be quickly taken without wasting much time at all.In addition, none of the team members need to wait to figure out the state of the regression. This is particularly helpful if tests are run in multiple sites and the results of the regression need to be consolidated by an engineer on the particular site. If an automated job is not put in place to update the regression results, then one has to wait for the engineer to update the results to everyone.This small oversight can be costly in projects with multiple sites and multiple timezones.

Automation of Status generation

Another key aspect of the regression summary is to be able to report automatically the status of the tests in the regression.Many tools like [LSF] do this automatically by optionally sending an email to the user when the test is done. However, what is important is that the status generated by parsing the log files and determining a pass/fail criteria is done automatically as far as possible. As indicated, the important thing is to determine as quickly as possible the results from the regression.

Label everything and use version control.

Almost all ASIC development today deploys a version control system. The Version control system like CVS [4] or Clearcase [5] or RCS or SCCS.

Almost all these systems provide a mechanism to mark a group of files with a certain unique name. This marking is created without actually creating a new version of the file.This process is called labeling of a file. The other alternative to this is to record the version numbers of the file used for verification. However tracking version numbers when there is a large amount of files quickly becomes an arduous task!

At the risk of sounding paranoid, one of the main things the author learnt from a senior verification engineer was that every file in the file system that was being worked upon and exchanged between teams should have a label on it. The RTL designer works on the RTL and then releases a label which represents a

[4] CVS is available from It is free software.

[5] Clearcase is a product of Atria Technologies. (as known to the author at the time of this writing) The trademarks and copyrights are owned by the respective owners.

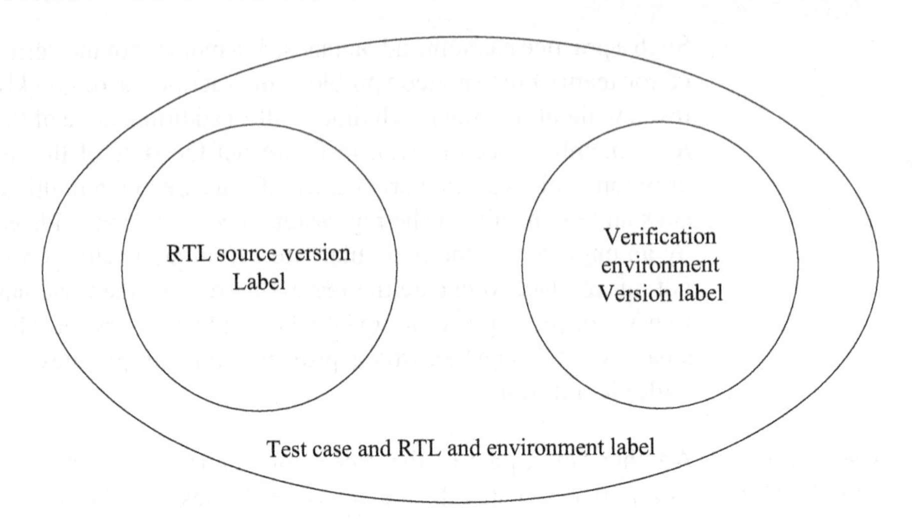

Figure 7.10. Use of Labels in a Regression

collection of files of certain version numbers as a configuration. The verification engineer accepts the RTL label and verifies it and responds with a label which is a configuration of files from the simulation environment. This label is a superset of the label released by the RTL engineer. Such a relationship is shown in figure 7.10.

This simple philosophy has saved endless hours when something critical needs to be reproduced. (see the case study on the *Effective use of source control strategy*).

Why is this concept of labeling a file so important? Using a label allows the user to take a snapshot of the code database and report all the issues with that version of code so that they may be fixed. It is entirely possible that the verification team may be able to find multiple problems with a certain version of RTL. Without a label or a version number, traceability is non existent. Labels are basically free. Once the label is very old and the RTL is mature,the older labels may easily be purged from the source control database using some administrative commands.

7.10.1 Identify Failures Before You Run Again

As the project approaches the end, the cost of running a regression typically rises. This is usually due to the fact that the regression at the end of a project has many many tests that need top be run. Every time a regression is run and the regression is broken, there is a test debug cycle that is activated. As can be seen from figure 3.3, the level of integration also implies that the size of the design is very large. Hence, each regression iteration becomes costly in terms of time and effort. Unless a collection of regression failures is completely understood, the author recommends that the regression not be launched again.

7.10.2 Don't Postpone Features to be Tested

This practice can have some undesired and unwanted side effects. In some cases some features may have some effect on the environment and on the test cases.The environment may need updating triggering frequent regressions that have no relationship with the RTL. In addition, some test cases may need to be updated as well.

Consider an example where the device had multiple clock modes. During the building of the environment, the clock mode support was not built into the testbench. Many tests in this example need to be updated as a result. If all went well, this clock mode support should not be a big issue. However, if the tests need to be repeatedly debugged since the clock mode support was not thought of beforehand, considerable rework and time loss occurs with no apparent progress overall.

7.10.3 Compile your Code

Manymodern HVL languages today have an option to use either an interactive or a compiled mode. The compiled mode usually has many debug features turned off and runs quicker during the regression. In addition, loading of several modules that are static when verifying some blocks may take a lot of time. using compiled libraries will help under these circumstances.

7.11 QC Processes to run a Clean Run

Prior to launching a new regression, some care needs to be taken to ensure that progress is made in a orderly fashion.

1. Bugs reported from the last defined label must be fixed and addressed. This makes sure that there are no pending issues from the current label.

2. The smoke tests for the new RTL release must also be clean with no issues.

3. All known issues from the RTL and test cases are identified and addressed before the regression is launched.

4. The previous run must be complete. This will allow regression management to be easily manageable without confusion in results.

7.12 Using a Data Profile to Speed up Simulations

Users running a simulation usually care about a few things.

1. How fast can this simulation be run?

2. Did the test pass or fail?

3. If the test failed, can it be debugged with what is available?

4. Where is the disk space to dump all this data?

A quick review of the data profile on a machine reveals a great deal. Some simulations generate a lot of data and it is often a challenge to manage this data. In addition to all this, network characteristics and performance play a key and important role in defining the way the overall simulation environment responds. Fortunately, more and more attention is being paid to this topic.

If we study the diagram in figure 7.11, this figure embodies a typical network architecture found in many organizations. In this example, there are a number of compute servers connecting to a central fileserver through a network. This network is

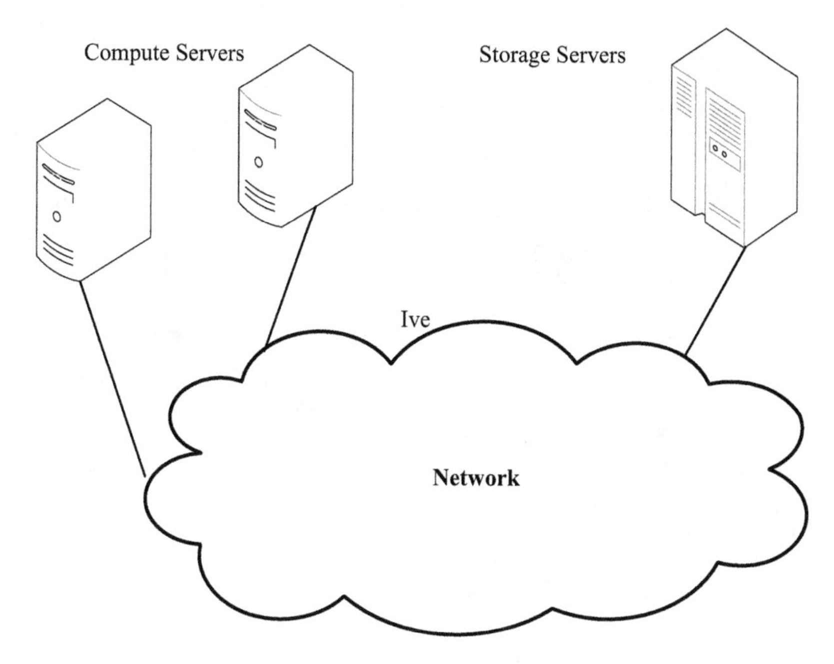

Figure 7.11. Data Dumping using a Network Drive

represented as a cloud since there could be several elements in it.

Assuming that the design in question is now generating a large amount of data, the data now has to make its way from the machine's CPU where it was generated on to the network and navigate to the server where it will be stored. If the data set is huge and is being generated fairly rapidly, then the slowest link in the chain affects the speed at which the simulation will run. In case of a 100 Mb/ps network would necessarily mean a 12.5 MB/sec connection between the compute server and the storage server.

An alternate view is to locate the data generated on the compute server locally and then send the results to the server at the time the simulation ends. The idea here is to communicate only the test results and the data if needed. If the test has passed, there is a probability that the data was not needed for the most part anyway. The information that is stored on the scratch disk can be very easily cleaned up by the post processing script that runs

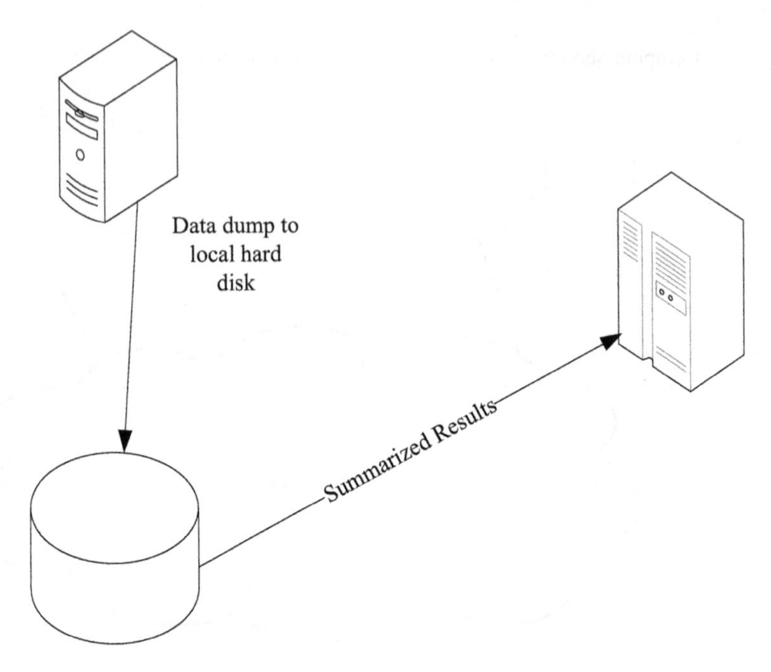

Figure 7.12. Dumping Data using a Local Disk

at the end of the test or by means of a script run periodically that acts as a data retention policy manager.

One of the considerations that needs to be taken when this approach is followed is the fact that the automounter[6] needs to be set up a little more carefully than the previous method. The author notes that this is not an impossible task since he has had the luxury of being in some environments where it was set up to operate seamlessly.

The network setup for this approach is a little more involved than the previous method. However, the connection between the CPU and the local hard drive is a local connection. The author has observed a 20% boost in simulation performance through this method when dealing with large amounts of data during debug sessions.

[6]The automounter is a program available on UNIX and UNIX-Like systems, It allows the mounting and un-mounting of network resources automatically.

Why is the concept of looking at local disks of any relevance? Fact is, centrally administered disk space is usually more expensive to start up and harder to get. They are always in short supply no matter which organization I've been at. On the other hand, most of the data is temporally close in nature and has a short lifetime.Once the information is gleaned from the data, the data is typically regenerated on another test run. Local disks make sense because one can get 100 – 200 GB of data for a very small sum of money [7] plugged into the local machine. Multiplying this by the number of machines in the regression environment, one notices that there is a huge amount of temporary space that is available for running regressions at almost no cost at all. This is particularly true with operating systems like LINUX. Since the operating system typically takes less than 10 GB of space and the minimum size of the disks nowadays exceeds 40-80 GB, this practice has an effect of speeding up data intensive simulations as well. It must be noted that the above discussion implies that there is some sort of a data management policy which periodically cleans up disks of outdated files. In the past, the author has found this approach very effective [8].

[7] The cost/megabyte of data has been falling for a while now. It is possible to get fairly large disks (160GB) for as little as 100$ US at the time of writing.

[8] To determine if this particular situation affects the reader's simulation environment, the reader merely has to run a data intensive simulation on a typical machine using the simulation scripts and the environment. The target of the data is set to be the fileserver. The total time taken to run the simulation is measured. This time is called T1.
The next step is to run the same tests with the same parameters on the same machine with the target of the data to be the local hard drive in a directory on the compute machine. The total time taken to run the simulation is measured. This time is called T2. If T2 is less than T1 by a threshold (say 10-20%) or so, then the organization will benefit from running the tests locally and using some sort of a script to report the results. This method works particularly well when there are large waveforms for debugging. Most of this data is scratch data which is useless after the data is looked at and the problem is understood. This method also works well if the temporary data generated during a simulation is large.

7.13 Getting the Machine to Document for You

In most verification environments, the test plan contains a list of tests and a plan of record revealing the tests being developed to test the device.

Usually a gap exists between what is documented and what is coded

Often, There is a gap between the tests being developed and the test documentation. The two are developed and completed at different times. This approach sometimes has problems. Sometimes if the gap is large, some details may be missed out during documentation and critical details are sometimes masked accidentally.

Usually, the author has found that the test plan documentation and the tests are sometimes uncoordinated due to paucity of time on the part of the test writer or for other valid reasons. Almost always, the paucity of time and the lack of importance given to documentation when the code is being developed lead to this scenario. A lot of time is sometimes wasted on document formatting and writing which can be minimized using some sort of templates.

Use scripts to cut the amount of work you do

In such a situation, if there was a way to cut out the work involved in documenting and keep all the vital information on record, then that system would be ideal. This system is described as below, as it has been effectively used by many people and is definitely in use by the software industry.

Place the documentation in the test

One of the key things that helps keep people organized is that the information that is required is kept in a single place. This also keeps things simple.

The cost of doing so is very low

The other key aspect is that the test code typically takes some time to mature and be final.

Looking at a test, for example, the test may take about a day or so to be developed, run against the RTL debugged and then saved as a final version. During all this time, the test file may undergo several modifications.

On the other hand, once the test debug is completed and finalized, the test itself contains a step –by – step description of what needs to be done to create and test the specific scenario. This can be easily placed in the test using some marked- up headers in the comment section of the test.

As can be seen from the figure above, a typical test case has the following information:

1. A copyright notice that states the copyright owners' name and conditions.
2. Test Name: Name of the test. Usually the file name.
3. Test Intent : what is intended to be tested in this test.
4. Test Description: How does the test go about achieving this?
5. Test Assumptions: Any assumptions that the test writer made.
6. Test Results: The kind of results that are expected if the test passes or fails.

All that the test writer has to do is to ensure that the description at the top of the test file matched with the description in the test. No additional care is required. A simple script can then harvest everything from the section TEST_DOCUMENTATION BEGIN and END sections and then format it according to the style that is prevalent in the organization.

This process is shown diagrammatically in figure 7.14.

This approach has a few advantages. First of all the test code and the documentation stay together. There is no opportunity for one or the other to be out of versions if the person updating the test updates the description at the same time. Since it is the same file, it is a trivial operation with little overhead.

All the relevant information is now in a single place

A reviewer can also look at the test and the code in one file and determine if anything needs to be done.All the information is one self contained file. Following the object oriented methodology in Section 3, the reader will be able to observe that the process becomes easy since the reviewer has to deal only with

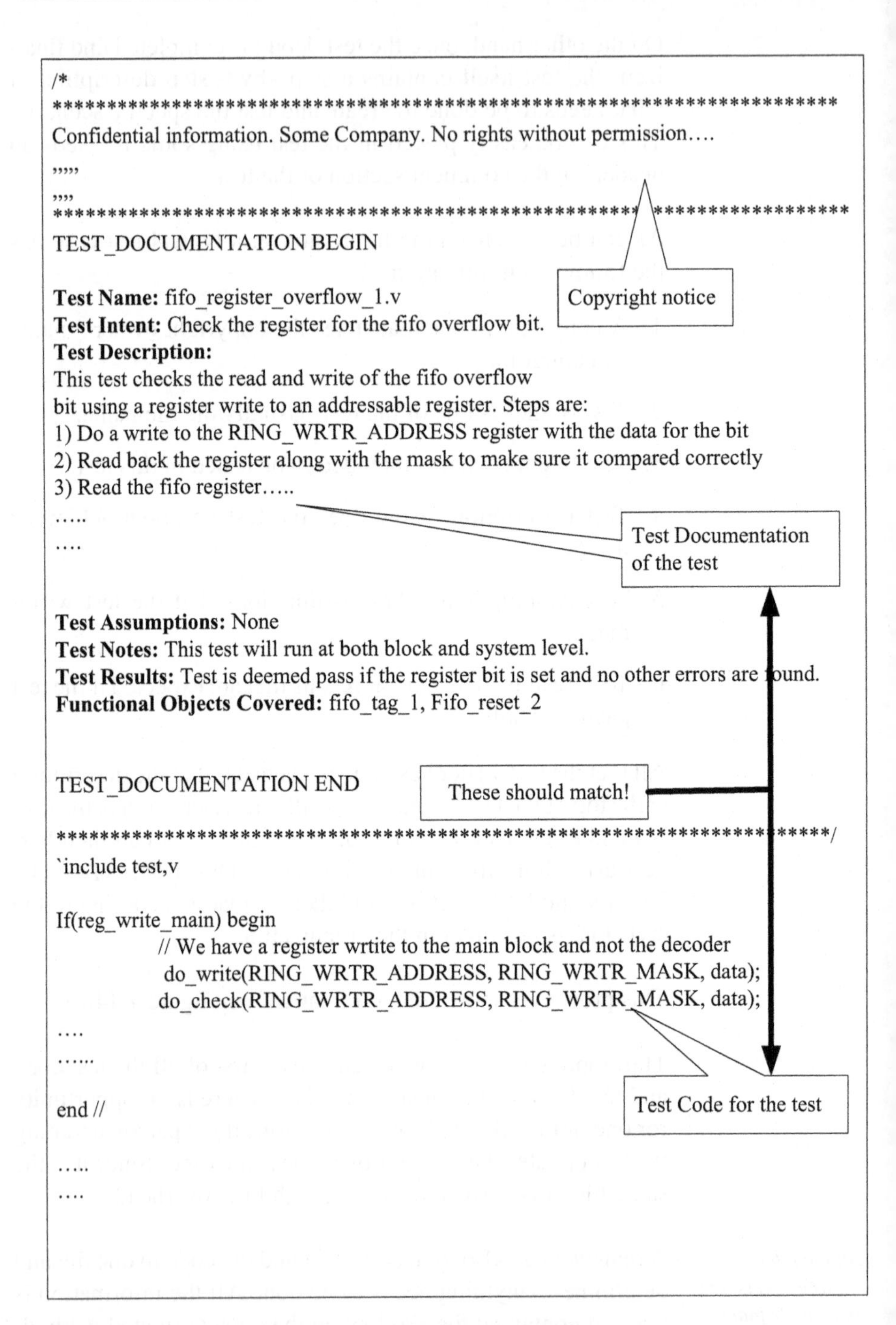
/*
****
Confidential information. Some Company. No rights without permission….
,,,,,
,,,,
****
TEST_DOCUMENTATION BEGIN
Test Name: fifo_register_overflow_1.v
Test Intent: Check the register for the fifo overflow bit.
Test Description:
This test checks the read and write of the fifo overflow
bit using a register write to an addressable register. Steps are:
1) Do a write to the RING_WRTR_ADDRESS register with the data for the bit
2) Read back the register along with the mask to make sure it compared correctly
3) Read the fifo register…..
…..
….
Test Assumptions: None
Test Notes: This test will run at both block and system level.
Test Results: Test is deemed pass if the register bit is set and no other errors are found.
Functional Objects Covered: fifo_tag_1, Fifo_reset_2
TEST_DOCUMENTATION END
**/
`include test,v
If(reg_write_main) begin
// We have a register wrtite to the main block and not the decoder
do_write(RING_WRTR_ADDRESS, RING_WRTR_MASK, data);
do_check(RING_WRTR_ADDRESS, RING_WRTR_MASK, data);
….
……
end //
…..
….
Copyright notice
Test Documentation of the test
These should match!
Test Code for the test

Figure 7.13. Test Documentation

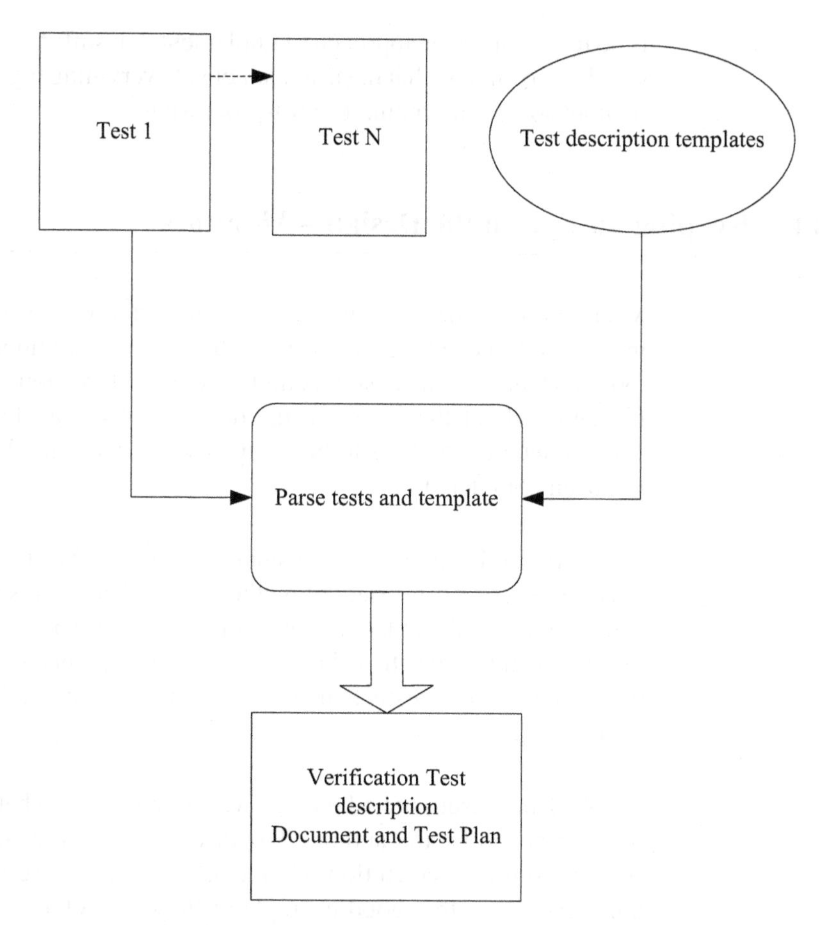

Figure 7.14. Process of using Test Documentation Automatically

a small number of files on a periodic basis.

Data can be used to correlate with other metrics

In the figure above, there is an entry for functional objects in the documentation portion of the test. This entry is a list of functional object tags that are associated with this particular test. How this ties into the overall picture is presented in the "putting it all together" chapter

There are other ways of implementing this concept

This approach to automatic documentation relies on a header script which does a post processing to create a document for review. On the other hand, many users the author has known simply "expose" the latest version of the tests using some sort of a filter to a web server and configure their browsers!! This is no doubt an easier approach to implement, but suffers from a small disadvantage that the user cannot get everything together without some fair amount of post-processing.

7.14 Keeping an Eye on the Design – Monitors

Monitors are used to observe the design and print some warning messages if needed into a log file. They serve as a important tool in debugging the design completely since they capture information about the events in the design into a log file and allow a detailed analysis to be completed after the simulation has been completed.

Monitors can be created and instantiated into the verification environment using a variety of methods. Some engineers have chosen to use HVL's like Vera or *e* which offer ease of use. On the other hand, some others have chosen to implement the monitors using various testbench constructs available in the HDL itself.

Some of the discussions here are written based on what has been practiced by the author. The discussions below is presented more as a collection of thoughts which may serve as guidelines. Another good example of these guidelines could be found in [34] which can offer excellent insight into this topic There are no doubt some good discussions that can be available in [1] or elsewhere in industry publications.

Each monitor has a single job

The monitor instantiated must observe all transactions on the interface.One important aspect of the monitor is that it should be able to decipher complex protocols if needed.This aspect of the monitor will save considerable time during debug if a failure were to occur. This approach is exemplified in the next chapter where the derivation of various monitors is made as part of the planning process.

Ensure that a module is a self contained unit.

Every monitor that is developed should not be relying on another checker or monitor for its operation. The code for the monitor should be self contained and complete in all aspects.

Why is this important? Having dependencies between the monitors not only makes the problem a little harder to debug, it also makes it difficult to turn off certain monitors and leave others on.For example: consider a certain monitor producing excessive amounts of output that are not entirely required for a certain simulation. If other monitors depend on the code in the monitor, it will be impossible to switch off the monitor and expect monitors with code dependencies to work. This leads to 'all or nothing' situation. – definitely not a nice situation to be in.

Later on in this section, each monitor is associated with a test object. This approach ensures that the monitor is completely self contained.

Any signals sampled by a monitor must survive the synthesis process

An important requirement for any signal that is indeed used by a monitor should be available after synthesis. Usually, What typically happens is that that there is the chance that the monitor will be unusable by the GATE netlists since the signal that the monitor was depending on no longer exists. As a result, it may become necessary to rewrite the monitor or retire it from service in the verification environment.

Writing the monitor to use interface signal boundaries is sometimes a safe bet and sometimes not. Many synthesis tools offer an option to keep module hierarchies.If a netlist is flattened, then all hierarchy information is lost and it may become very difficult to hook up the monitor.

Disastrous results loom if the monitor is written after tapping some signals in RTL without paying any regard to whether the signal will indeed be available after the synthesis process

Any unusual interface behavior ought to be flagged as an error by the monitor

Another important aspect of monitor development is that the monitor must be in position to display a message of some sort when it sees a behavior that it does not recognize. For example, if a monitor is set to observe some transactions on a read only bus, and observes a transaction that did not look like a read transaction, it must flag an error.

Use debug levels for the monitor. This can be used with various levels of debugging

Every monitor should be developed keeping in mind that there are certain instances where extensive monitor output is redundant or no longer needed. It must be possible to turn off the monitor if required. When the monitor is being developed, the author recommends that the monitor be coded with multiple debug levels. Having multiple levels of debug allows the user of the verification environment to customize the verbosity of a certain monitor depending on the needs of the debug effort.

Use Monitor output to do data analysis and reporting

One of the important features of a monitor is the ability to analyze the data available from the monitor after the simulation is done. For example, considering a bus protocol where there are several transactions between several devices, the monitor output could be post processed to observe the latency of the bus under certain conditions. The output of the monitor could also be used to keep a running trace of the kinds of scenarios experienced during the simulation. In the author's opinion, the fact that a monitor output could be used for other purposes is considered a free benefit of having the monitor.

7.15 Checkers in an Environment

Ensure that a single checker does a single job

One of the main guidelines when developing a checker or a monitor is that it's parameters are well defined. It must be designed to do a single job. For example: the author has myself written a datapath checker and merged it with a protocol checker. When the protocol went through some variations, it became difficult to maintain both the checkers at the same time in the same file. This for me led to grief sooner or later.

For example:If a checker is designed to check the protocol, it is crucial to let the checker focus on that aspect alone. Latency and data path checks should never be made part of that particular checker. The requirements for a latency checker are quite different from those of a datapath checker. Mixing the two requirements will invariably result in a checker that does a little of both.

Often, attempt is made to create a single checker that will do it all. The author had been in that situation himself! Two different types of checkers were embedded into a single checker module. The end result was that one of the checker seemed to work well and the other was seriously impaired. The faulty checker was causing havoc with regressions. Since the two checkers were clubbed together, the decision was taken to rewrite them as separate units after effectively throwing away the effort built into the checker.

Checker specification should include the ability to recover from errors

Recovery of state machines in the checker should be part of a checker specification. Any well written checker should be written in a way to help the checker recover from errors. Errors may be from multiple sources. The design may be in an erroneous state caused by some test case or data. The checker-may also be forced into a unique state since it is observing the data patterns passing through the device.

Being able to recover from these errors is essential for the checker, it should not need a reset of the design and the environment to be able to reset the checker. One of the most catastrophic things that can happen is that the design is able to handle abnormal inputs and the checker aborts the simulation.

Make the checker self contained.

A key attribute of a checker is that it must be self contained. All the necessary information for the checker should be contained within itself. In many cases the checker may choose to instantiate some existing monitor code to promote reuse. However, the same should not need to instantiate another checker for completeness.

Such a guideline is essential to promoting reuse of the checker in other environments. For example: Let us consider a checker A that instantiates another checker B for its proper operation. In another derivative design, the features offered by checker B may be redundant or not of vital importance to the specific design. However, features offered by checker A may be of vital importance. Since the checker B is now a part of checker A, checker A will effectively have to be rewritten although not much of it has probably changed!

Be careful with clock domains.

One of the important aspects of placing a checker on a single interface is that it allows the checker to be completely self contained as far as a single interface is concerned.

Clock domain synchronization is one of the more trickier aspects of digital designs. Many designs devote time and energy to ensuring that the design operates correctly when clock domain crossings are present.

A similar approach needs to be taken for the checker code as well. In any design with multiple clocks, it is important to ensure that the checker either operates in a single clock domain or is designed to handle the data across multiple clock domains correctly. Not paying attention to detail may cause the checker to fail in some specific scenarios and expend significant resources of both the design and verification engineers to find the source of the problem.

use monitors or debug statements inside checkers if needed for advanced debugging of the checker

Many a time, a checker and a monitor are built together. The author recommends that the monitor be built separately and instantiated in the checker if need be. Adopting this practice may at first seem to make the checker design ungainly and unwieldy. However, if a monitor is used in multiple places, then any bugs found in the monitor and corrected will also benefit the checker. The added advantage is that there is proper reuse of the modules in the verification environment.

It is considered it a good coding style to use existing monitors wherever possible and enhance the debug capabilities of the monitors using multiple debug levels.

Do it right and do it once!

In a design the author worked on, a checker for a collection of modules was worked on by a team. It was buggy and incomplete at best. It used to fail frequently and sometimes for the wrong reasons as well. Many false failures were reported and time lost over it. Over a period of time, it was taken out of the environment rather than fixing it.

7.16 Linting Code

Some engineers who develop their code tend to ensure that the RTL is mostly lint free before they actually submit the code for verification. This approach has ensured that the most common errors are caught up front and the verification can focus on other important things. (See the chapter *Cost of doing things incorrectly for an actual case study).*

7.17 The RTL Acceptance Criterion

The verification team is typically a separate and independent team from the RTL. Depending on the size of the team and the degree of cooperation between the teams, Varying degrees of formalism in accepting RTL for verification have been observed by the author at various companies.

In some organizations, the module owner typically creates a module level testbench and verifies the code to some degree before handing it off to the verification team. In others, the RTL designer writes the code, makes sure it compiles and without any verification whatsoever "throws it over the wall" to the verification engineer for verification. This kind of a dramatic range in RTL quality brings about a few criteria for accepting RTL as discussed below.

In some cases, the verification suite is extensive. During the later cycles of the project, the verification engineer can insist that the RTL pass all the tests that passed on a previous build of the module prior to submitting the RTL for verification.

In some other scenarios, the verification team makes sure that the module submitted for verification at least passes tests designed to ensure that there is at least a minimum quality before accepting the module for verification (SMOKE tests).

Why is an acceptance criteria even required? Is not the verification engineer's job to make sure the RTL is of good quality? The answer lies in the fact that the verification activity is usually a much bigger activity in scope than the actual design activity. Making sure that the RTL is of some good quality helps the verification engineer to concentrate on activities that add more value instead of dealing with simple problems that cost a lot of time and effort. Similar thoughts are shared in [1] In order to do this however; a good deal of teamwork and respect for one another is required of both the RTL and verification teams. One interesting thing the author has observed over the years is that RTL designers who submit good quality pre-verified code for verification have always earned the greatest respect from their verification peers!

Conclusions

This chapter presented various techniques to reduce work in verification. As the reader is aware, many of these techniques are common techniques in the industry. The appendices provide detailed information to enable implementation of the concepts presented herein. It is hoped that the reader is able to implement some of the presented concepts and is able to take verification productivity to the next level in the project that they are working on.

References and Additional reading

[1] Todd Austin. Building buggy chips that work! *Presentation from the Advanced Computer Lab*, 2001.

[2] Kudlugi, Murali, Hassoun, Soha, Selvidge, Charles, and Pryor, Duaine (2001). A transaction-based unified simulation/emulation architecture for functional verification. In *DAC '01: Proceedings of the 38th conference on Design automation*, pages 623–628, New York, NY, USA. ACM Press.

[3] Bening, Lionel and Foster, Harry (c2001). *Principles of verifiable RTL design - a functional coding style supporting verification processes in Verilog*. Kluwer Academic Publishers, Boston, 2nd ed edition.

[4] Kuhn, T., Oppold, T., Schulz-Key, C., Winterholer, M., Rosenstiel, W., Edwards, M., and Kashai, Y. (2001). Object oriented hardware synthesis and verification. In *ISSS '01: Proceedings of the 14th international symposium on Systems synthesis*, pages 189–194, New York, NY, USA. ACM Press.

[5] Bergeron, Janick (2003). *Writing testbenches - functional verification of HDL models*. Kluwer Academic Publishers, Boston, 2nd ed edition.

[6] Motorola (2003). *The functional verification standard.*

PART IV

PUTTING IT ALL TOGETHER

This part describes ten steps to help the reader get started from a verification plan to successful silicon. The steps presented to the reader are design and tool agnostic and discuss basic principles that can be used to verify any design. The procedure described does combine some concepts discussed in earlier parts.There is a single chapter in this part titled *Putting it All Together.*

It starts out by helping the reader identify the features of the device that need to be verified. Guidelines are offered to the reader to help complete a review of features and identify a test case list. The discussion of various types of test cases and test strategies is presented here.

The chapter then goes on to discuss the concept of a Test Graph Matrix which is a grid like structure allowing the user to apply high level optimizations to the verification effort. These optimizations allow the reader to become effective at generating high yeild test cases.

Being able to track where one is in the overall verification effort is crucial to the success of verification. This chapter builds on concepts presented in the previous part by showing the reader how to embed "tags" that simplify the task of managing the verification effectively. The chapter then concludes by discussing the need for Gate Level simulations and the steps in verification signoff.

Chapter 8

PUTTING IT ALL TOGETHER

Ten Steps to Success

We have discussed several aspects of functional verification in the last few sections. Now the task ahead is to bring all these aspects into a functionally complete, executable plan that helps assure good first quality silicon.

A good functional verification plan is important to success of any project. While most managers wind up asking the question, "*when are we done?*" in turn, the engineers reply "*We'll be done when we're done!*"

Being able to tape out on time and answer the above two questions is one of the crucial aspects of a good verification plan. A good plan helps engineers to address all concerns with the device's functionality while allowing adequate visibility to the program managers without excessive reporting overhead.

Another key aspect of verification plans is that it should be able to predict accurately the amount of effort required to tape out the device.This is indeed a difficult question. Hence, many teams use a verification plan that is defined to be as complete as possible and then attempt to tape-out the device based on certain commonly used metrics. These metrics were described in the *Tracking Results That Matter* chapter.

Nowadays the world has become a "global village" as far as ASIC development is concerned.The design is conceived in one part of the world, built and probably verified in a different part of the world. Hence, it becomes vital to use a process that allows various teams to exchange information in a smooth manner.

On further reflection, it becomes apparent that no one solution fits various kinds of designs in the marketplace. Given that design complexity is rising with increasing levels of integration, an effective object oriented methodology can be very useful in helping manage the challenge posed to verification teams to completely verify the device under verification [1].

Verification has traditionally been an ad hoc process where the specification is reviewed and tests are written to match the specification. In some organizations, a well defined process is used to derive the tests that need to be done for verification. In addition, a variety of factors affect the verification challenge. These factors exist in varying degrees in different organizations as is evident in the following paragraphs.

Existing infrastructure problems

It may not be possible to run all the appropriate test cases on all process corners due to limitations in infrastructure. While some of these may be solved easily, under some circumstances due to the size of the design, it may require a move from a 32 bit platform to a 64 bit platform. During such a move, many scripts and other programs may become inoperable further compounding the issue.

Legacy design constraints

In some cases, the design may have been taped out due to various reasons and may not have been completely verified. Hence, the tape-out would have some known and some unknown bugs. Later, it would be difficult to make changes to this design since it is deemed to be "working" albeit with some known issues.

Human factors and design expertise

The individual who had been working on the design may not be presently assigned to work on the design. Alternately, the person may have chosen to part with the organization. It is also possible that the team member currently assigned to the verification effort may not be well versed in the design and may need significant time to acquire knowledge of the design.

Schedule constraints

Schedule has always been the adversary of verification activities. Many a time, a tight schedule may cause significant pressure on verification activities.

Cost constraints

Under some circumstances, the verification may be operating with a fixed budget due to various other factors that may be at play. As an example: for cost reasons, portions of the verification may be outsourced and done by a third company. Alternately, there may be a limited number of licenses available, and it may not be possible to obtain all the required licenses all the time. This may result in longer turnaround times to complete regressions.

Physical constraints

In some of the contemporary designs, the design and verification teams are sometimes in different locations and in different time zones. This adds to the difficulty for communications and other debug activity that may take place between design and verification engineers.

There is no one solution to this challenge posed to design and verification engineers. Different problems demand different approaches. However, the core principle behind all methodologies is ensuing that the device will work as specified.

The above discussion makes it apparent that the entire verification challenge is a multi faceted one which involving may items in several disciplines coming together to make verification a success. Many scripts in possibly multiple languages, approaches, and strategies typically find their place in the verification approach. The entire effort can be looked upon as a puzzle. Some of the main pieces of this puzzle are depicted in the figure 8.19. Many of the items in the puzzle owe their roots to the verification plan. This plan needs to be comprehensive in all aspects. Hence, a step by step approach to ensuring completeness and correctness is essential.

This chapter is all about building a successful test plan and ensuring that it stays the test of time and the course. The methods described herein can be adopted in part or in full based on the particular scenario and situation. It is designed to grow with the user's needs and can be customized fairly simply. The process presented is a synthesis of much iteration and attempts to

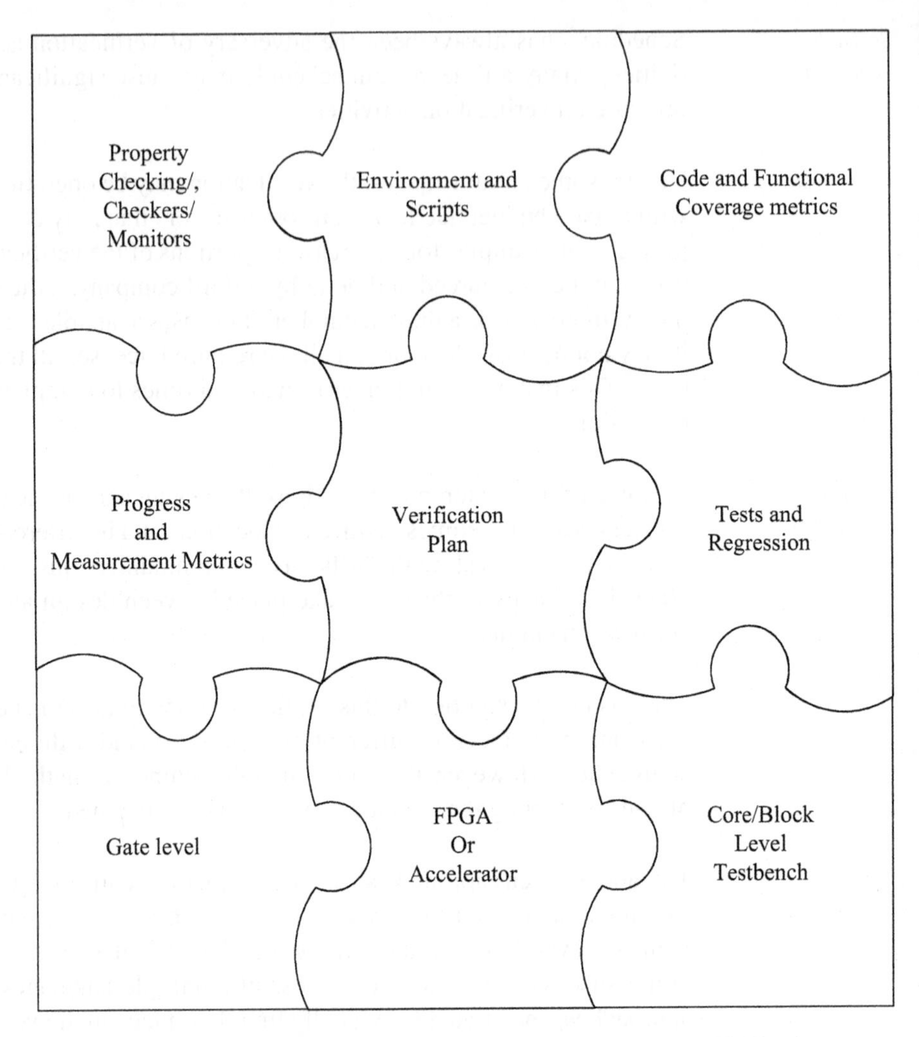

Figure 8.1. Components in the Verification Effort for Pre-silicon Verification

address the verification challenges listed above using a step by step approach.

As can be observed from the figure in figure 8.19 , several aspects of verification need to work "hand in glove" with one another in order to successfully complete the device validation.

Verification Plan

The verification plan is one of the most crucial aspects of successful functional verification. The verification plan includes tests to verify the device at various levels of integration. This plan may be supplemented by various methodology documents and other documents which specify the scope of the verification effort.

Assertions/Monitors and Checkers

The also are a part of the implementation of the verification plan. These are used to monitor the health of the device under verification. Various considerations for these components were presented in the earlier chapter. The derivation of the assertions, monitors and checkers is covered later on in this chapter.

Environment and scripts

The environment for verification must be clean and easy to use and various scripts in the environment must be able to execute on the appropriate platforms. The implementation of the scripts usually are done in a scripting language like PERL or TCL. There are many excellent resources on the web on the topic of script development and the reader is referred to them.

Code and Functional Coverage Metrics

The discussion of various metrics was presented in the chapter *Tracking Results That Matter*. The metrics provide a means to measure the extent to which the device has been verified.

Progress and Measurement Metrics

The are typically the Bug Find/close rates, test object density, and other metrics presented in the chapter *Tracking Results That Matter*. These metrics provide a mechanism to reveal the progress in verification of the device.

Tests and Regression

The are always key components of the verification effort. Effective selection of tests and their various types is discussed in this chapter.

Gate Level

Gate level regressions with both zero/unit delays and back-annotated SDF are essential to ensuring that the design will operate properly at various temperature and voltage corners.

FPGA

FPGA are also a useful method to verify that the design is indeed functional before a Tape-Out is attempted.

Core/Block Testbenches

Various aspects of the testbench and other components was described in the earlier chapters. It is essential that the testbench be extremely efficient in order to allow the simulator to spend the maximum time simulating the device under verification. Some considerations were discussed in the earlier chapters.

It is hoped that the step by step approach that is discussed allows the reader to identify ***the items that are important*** rather than ***how to develop tests and other items in the verification environment.*** The latter is actually well covered in many texts[2], [3] etc. While the latter consumes most of the time, Identifying the former is usually the bulk of the work in verification and most challenging. Once the scenarios and other items are identified, it becomes a straightforward task to implement and test the device.

Verification is in the author's opinion 70% methodology, planning and strategy and about 30% work. If planned correctly, the verification experience will be a very enjoyable one.

The approach in this book has been to keep any implementation specific discussions outside the realm of this book and focus on concepts. This approach is deliberately chosen to enable the reader to come up with an implementation easily to suit the needs of the particular situation.

The implementation described is a generic implementation that can be adopted easily

The steps presented have come about as a synthesis of working in many companies. While it would be ideal to expect the reader to follow all the steps in sequence, The author realizes that each verification engineer has their own personal style to solve the verification challenge. It is hoped that the steps presented could also offer the reader some ideas to augment what the reader already has in the verification environment currently under development.

The way the steps are laid out, the approach appears to lends itself very well to situations where the design is starting from scratch. This is many a time not true. However, in a legacy environment, the author has used this approach to help build correlations, feature lists etc effortlessly toward the end of many projects to help ensure verification quality. In one instance, several changes were made to the tests and environments to overcome shortcomings and bugs in the design exposed by the process described herein. It may be possible to follow the steps for only portions of the design that have changed. The author encourages the reader to use whatever steps are appropriate in his/her environment.

There are no doubt many other approaches as evidenced by literature [4],[5],[6]. However, given the nature of the verification process, The reader will discover that that the approach represents a significant reduction in the quantum of work both from a workload and management point of view.

Habit 1 described the concept of keeping the big picture in mind at all times. Hence, the data for the verification plan is presented in a "grid" like format, so that the reader can get a complete view of the verification activity. The reader may choose to implement the plan using a design of the reader's choosing without using the grid like structure presented in the sections that follow. It may be possible to store the results in a standard word processing document by merely generating the data from the grid into the document and reviewing the document for completeness and accuracy[1]. The effectiveness of the process is not reduced using any particular data format as long as the principles are followed.

A bird's eye view of the entire process is presented in the figure 8.2. This process should be self explanatory for the most part. The reader is encouraged to refer to this figure as various steps are presented in the sections that follow.

8.1 Step 1: A Specification Review

"The journey of a thousand miles begins with the first step" –old proverb

Prior to beginning a verification effort, the top level architectural and implementation specifications should be finalized or as close to final as possible. This is one of the starting steps in building a solid verification plan. The top-level architectural descriptions provide a view of the device behavior while the implementation specifications provide insight into the specifics of the device under test. The device being designed may be required to conform to certain specific industry standards. These standards may also specify specific responses that the device is supposed to have for certain conditions.

[1]The implementation of this concept is given in the appendices.

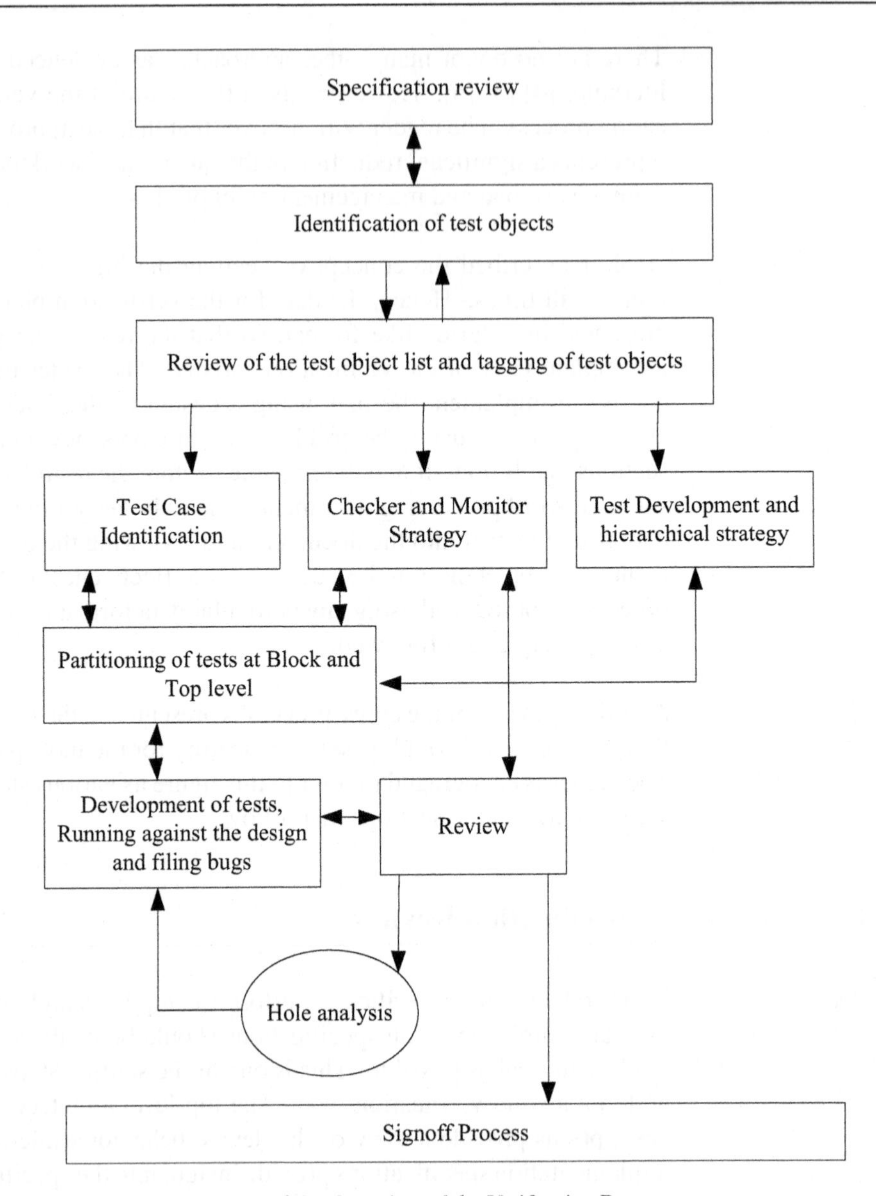

Figure 8.2. Overview of the Verification Process

A complete understanding of the specifications and the target socket for the device to be verified is a manifestation of ***Habit 1.*** Understanding of the specifications is important to making sure that the device is verified correctly and operates as specified when it is fabricated on silicon.

Many a time, the specifications are not final

This happens all the time. Especially in some innovative products where product research mingles with product development, there is a chance that a significant specification "creep" happens during the project. The duty of the verification engineer is to make sure that the nature of the changes, the driving force behind the changes and its impact on the verification environment are understood. It is noted that the environment and tests need to be robust to be able to handle specification updates.

Verification effort can also influence the specification

In current times, modern System on Chip devices (SOC's) attempt to reuse portions of the design from other previously released designs. This common trend is seen today. One of the common observations that are made is that the size and complexity is growing for the current generation of ASIC's. If a certain portion is too expensive to verify, then the system architects can look into how the existing effort can be leveraged while being able to create newer generation devices without impacting verification extensively [2] [3].

[2]This actually happened on a couple of projects the author has worked on. In one of the projects, during the initial stages, we found that we could not implement and verify a design which was extensively using memories. On analyzing verification impact, the module was redesigned to eliminate the memories altogether.

[3]In another project that incorporated a processor like state machine that the author had the opportunity to be associated with, the design was altered to accommodate the fact that it was impossible to verify the design completely. The newer version of the design cut the verification effort significantly.

8.2 Step 2: The Identification of Test Objects

Extract features from the design that will be verified

The next step is to identify various features of the design.The features are usually extractable from the specifications of the device.A design today can be split up into various kinds of features.

Application level features

Application level features of the device typically are features that define the behavior of the device at the application level. For example, the device has a MPEG2 decoder, or a Bluetooth module.

Under normal circumstances, the verification engineer must be able to obtain specific information about the typical application of the design so that he can determine the various aspects that must be captured at the application level.

Application level features may take advantage of some behavior inside the device as well as outside the device.Hence, it becomes necessary to capture all relevant information regarding these features.

Application level features can involve multiple interactions across multiple interfaces, sequentially or concurrently. Because this type of coverage information often encompasses the entire design, or at least significant portions of the design. The application level features are typically captured at the subsystem or at the full chip level.

Interface level features

Interface level features are the most common form of functional objects. Internal-interface features are captured at the block and subsystem Levels while the external-interface features are captured at the chip level or at a multi-chip system level.

Since interfaces are both internal and external to the device, capturing interface protocol information occurs at the block, subsystem and chip levels of verification.

Examples of the interface include the various types of reads and writes across a bus, such as the peripheral component interconnect (PCI) bus, the advanced micro-controller bus architecture (AMBA) bus, or sending and receiving frames across an ethernet or SONET port.

Structural features

Structural features relate closely to the implementation of the design. The structural features may be found in the micro architecture of particular blocks of the design. The following are examples of design elements that embody structural features in the design. Some examples are:

- Finite state machines (FSMs).
- First-in-first-outs (FIFOs).
- Memory elements.
- Arbiters.
- Handshaking on an interface.

It is noted that some structural coverage is indeed offered by code coverage, namely state machines.

Specific implementation information

The design may have some specific design implementation which needs testing. For example, the use of a memory buffer with some upper and lower thresholds, the use of certain I/O signals in a specific way etc. Such information is usually available from the design documents.

Configuration information

The device may have a set of registers that may be used to configure the device in various modes. Some of the behaviour of the device may be changed by configuration. All the possible modes and possibilities will need to be tested.

Protocol implementation

The device may choose to implement all or part of a standard protocol. The compliance to the protocol will contain many items which will need to be verified.The list of these items is also added to the test object list.

I/O Ports of a module

The pins of the device may exercise some control on the functionality of the device. In addition, the device may accept data in a certain format and offer data in a certain format on the pins. The controllability and various modes and formats will also be test objects needing to be tested.

Industry standard specification

Many times, the device may have been required to have a specific response to certain inputs which have been defined as an industry standard. The compliance of the device to the standard may involve some standard checks which will be part of the test object list.

Performance of the device

Many devices also have a collection of performance goals to be met. These goals may be in the form of noise performance, bandwidth etc for some mixed signal devices or alternately throughput parameters etc. These performance requirements for the design may also need to be verified. Each of the performance requirements must also be added to the list of test objects if they must be verified.

This makes for a long list! In case of an SOC, the list can be huge. Looking at a large specification, one can easily become a little challenged on the size of this task. Abstraction and divide-and-conquer approaches [7] are some of the known tools known today to help reduce complexity.

One of the approaches frequently adopted to conquer the feature list completion is to request the appropriate engineers to actually share their lists of features for the block in question, (divide the chip into parts and involve the design and verification engineers for the block in question). The lists are then correlated and a master list is generated that acts as a superset of the lists provided by the engineers.

Since many people now work on their lists, the lists are easy to generate. Every verification engineer has a list of items to be tested anyways. What is being done is to build a top level list that can be used to understand the complete picture. Adding to the list user level scenarios and discussing these in a review (done in the next step) usually completes the list.

Each of the items on the list can then be broken into several atomic properties to add granularity. In the description moving forward, these are addressed using the term test objects[4]. At the end of this step, there must be a complete list of test objects

[4]For the purpose of the discussion that follows, the terms "test object list" and "feature list" are used interchangeably in this book.

for the design. This list is called as the test object list or a feature list moving forward.

Feature list can contain objects at many levels

Many a time, the author has been asked the question: *Does it really matter if some of the features are elaborated properly and some are at a little higher level during this process?*The fact of the matter is that it does not matter too much if you have a firm grip on the features that are being verified.

At some stage of the list development, the list may only contain high level verification goals. As verification progresses, It will become apparent that the high level verification goals will eventually have to be broken down to lower levels at some point during the verification process to allow the implementation to proceed. This natural process will happen as a course of evolution of the verification plan's development.

The test object list may contain some duplicates when the list is being created

The only effort that has to be made is to keep the list as complete as possible. At some point, there will be duplicates. Since the test object list is actually a collection of inputs from a variety of sources, it is possible that during the initial stages, the list will contain some duplicate entries. The list may contain some high level objects and some implementation specific ones that refer to the same feature.

Periodic reviews are part of any process. These reviews usually have the effect of consolidating duplicate objects and eliminating redundant ones. As long as the list can be maintained, it is suggested that the list be provided with a mechanism that allows certain entries to be invalidated while leaving the list intact. When the tape out review is undertaken, the list of invalid and duplicated entries can be carefully reviewed to ensure that there are no missing scenarios or link to tests that have been omitted from the verification effort.

Do the fact that there are duplicates mean that there is lot extra work? It must be apparent that each engineer is working on a portion of the list. Hence, duplicates show up immediately.

What has typically worked is to maintain an "invalid/reject" column on lists that allows one to keep track of objects that have been invalidated or replaced by others. During a review, what is normally done is to prove that the replaced item is indeed covered by other items.This approach has ensured that nothing is missed.

Change is about the only permanent thing in this universe

The list will be a working living document. It must be anticipated that during the course of the project, the list will probably grow in some areas.The fact that the list has changed, implies that one has to be in a position to know if the items on the list have indeed been replaced by newer entries that clarify the design intent a little better or is some new entry to the list.

Is list management a great deal of work? For the most part, the list once created and reviewed will only require periodic maintenance.

8.3 Step 3: Review of the Test Object List

In the previous section, A test object list that was as complete as possible was created. This activity might initially sound a little bit redundant. However, the list actually is the foundation stone of a successful verification strategy.

Identifying the specifics of the device is sometimes the difficult part of verification plan development. The intent is to make sure that the list of items to be tested is complete. This can be accomplished by reviewing the functional list of objects with the designer of the module or other verification members of the team. This is an important step that must be taken early on in the verification planning stages, this step ensures that the list of test objects is complete. Any functional requirements or objects that are missed out in this phase may have some larger implications later on

Another offshoot of this step that has been observed is that it helps clarify the functional behavior of the module or device under test to both the design and verification engineers. Any missing information from the specifications is quickly ironed out in this step.

It is not imperative to do the entire review in one sitting

Conducting a test objects review is a process that is sometimes known to take some time. Many engineers have hesitated to take this step, wanting to go "piecemeal". However, the value in this step is such that all parties have a complete and clear

Block Name	***Feature Object***	***Priority***
Cache	Cache arbitration for Port B read	**p1**
Cache	Cache arbitration for Port B write	**p2**
Cache	Cache arbitration for Port A read	**p1**
Cache	Cache arbitration for Port A write	**p1**
Cache	Cache flush algorithm	**p3**
Cache	Cache controller states	**p1**
Fifo	Functional soft reset	**p1**
Fifo	Insert 0 pattern	**p2**
Fifo	Autoincrement feature	**p3**
Fifo	Frame timing reference	**p1**
LLC	Different n bit positions for L1 and L2	**p2**
LLC	Continuous output pointer increment	**p2**
LLC	Payload crc insertion	**p1**

Figure 8.3. Features with a Priority Assigned to them

understanding of the task in front of them. It is not required that the entire review be conducted at one go, however, this step is a gating step to executing a well written test plan. This step must be completed before any further progress is made.

It now becomes possible to associate a verification priority with the feature list

In the figure 8.3, a list of features and their priority is illustrated. During the review of the test object list, it becomes possible to identify some of the key features of the device that are critical to the success of the device. Some of the other features may be "nice to have". Hence, a verification priority can be set on the features. The high priority features can be verified thoroughly with appropriate priorities if other factors like schedule/market pressure are at stake.

The test object list is NOT a test list by any means

Successful manifestation of ***Habit 2*** by verification teams helps ensure that this step yields a complete list of test objects available and allows the teams to complete the review quickly. Breaking the list of test objects into small chunks helps teams usually make this more manageable.

It must be noted that the test objects must not be confused for test cases or test scenarios. At this point, the main thing that must be captured is the functional features or properties of the design. The actual test cases and test scenarios are extracted a little further in this process.

8.4 Step 4: Tagging the List of Test Objects

The list of test objects was created in the previous step. In the steps that follow, this list will be used extensively both by humans and by computer programs. To help manage the information in a concise manner, it becomes imperative to give each feature a distinct name.

Why is naming of test objects even a consideration? One notices that there are features that are at various levels. It was mentioned before that the list may contain test objects at different levels. At some point, these get mapped into test cases.

Given that one needs to manage a list that has the potential to grow or shrink over a period, it becomes essentially obvious that some sort of method to identify objects in the list becomes necessary. This is accomplished by means of a unique string that uniquely identifies the test object in the test object list.

Information in databases all over the world use a similar conceptual idea by means of key or index to quickly get the information to the user.

In the sections that follow, this name is referred to as a "tag". This tag can be anything of the readers choosing. Even a serial number of the item in the list is acceptable for this step. However having an alphanumeric string of some sort that adds some description usually helps later on in the verification cycle. This is because looking at the name, one might be able to infer something about the particular feature in question.

The example in figure 8.4 indicates a list of tags chosen with a naming convention. The convention uses an '_' to separate groups of characters. The first group of characters indicates the block name, the second group of characters indicates the type of feature, and the third group indicates a particular module in the design and so on. In a similar manner, a naming convention of the user's choice can be used.

There are some advantages to tagging the test objects. It allows a computer program to help keep track of the objects. It also provides a short description that could possibly communicate

```
BLK1_FUNC_FIFO_FULL1                          -- Fifo is full
BLK1_FUNC_FIFO_EMPTY1
BLK1_FUNC_FIFO_HALF_FULL1
BLK2_FUNC_FIFO_HALF_FULL2
BLK4_FUNC_FIFO_HALF_EMPTY1
```

Figure 8.4. Use of Tags

a great deal effectively when some debug is needed during a regression. The other advantage is that it now becomes possible to use some filters or other pattern processing tools to act on the results of a test should such processing be required in a regression environment. An example of using the tags to remove tests from the regression for features that are known to be broken is shown in the next section.

In this chapter, the concept of tags is used extensively. The reason for using the tags is that it presents the user with a simple easy way to keep track of what is happening in the verification environment.

8.4.1 Using Tags to Simplify the Regression Process

Using the scheme above, one aspect of application of the tags would be to identify features that fail in a regression. For example: One may choose to filter out all failing tests that have anything to do with BLK1_FUNC_FIFO_EMPTY1 since it may be known that the particular function does not work at that point in time or has a problem in implementation.

An illustration is provided in the figure 8.5, Two different approaches are contrasted. The first is that of a regression that did not use the tags, The second is a implementation that uses the tags.

In the example, the jobs in the first instance were named using a script. Assume during the course of a regression, it is assumed that a discovery is made that the feature BLK_FIFO_1 and BLK_FIFO_FULL_2 don't quite work right and cause tests to fail.

The verification engineer would typically wind up killing the entire regression for that particular module and restarting it after the design engineer fixed it. It is safe to assume that the engineer would not go about finding those specific tests before he/she issued a kill command to the regression. Without a naming mechanism, it is just too much work.

On the other hand, using tags as a portion of the job name. the output from the queuing system would look similar with one key difference. The name of the job now uses the tag selected above as a portion of the job name.

When the regression is running, if we assume that all tests with the particular feature BLK_FIFO_1 fail, then all tests that test that particular feature still in the wait queues can be easily identified and killed. the rest of the regression can probably proceed to uncover other issues found with the design. The initial debug can start with identifying the specific circumstances that caused tests with BLK_FIFO_1 to fail by merely looking at the jobs

You cannot distinguish features without a little digging in!

JOBID	USER	STAT	QUEUE	FROM_HOST	EXEC_HOST	JOB_NAME	SUBMIT_TIME
3621	mendonca	RUN	short	jupiter.a.eng.r.com	pdsflx23.ne	psax	6/29/2005 11:32
3851	shiv_runall	RUN	short	ganymede.eng.r.com	pdsflx23.ne	job1_1	6/29/2005 10:30
3852	shiv_runall	RUN	short	ganymede.eng.r.com	pdsflx23.ne	job1_2	6/29/2005 10:31
3853	shiv_runall	RUN	short	ganymede.eng.r.com	pdsflx23.ne	job1_3	6/29/2005 10:31
4001	robjose	RUN	short	saturn.eng.r.com	pdsflx23.ne	regress_1	6/29/2005 12:31
4002	robjose	RUN	short	saturn.eng.r.com	pdsflx23.ne	regress_2	6/29/2005 12:31
4040	ribbit	RUN	short	saturn.eng.r.com	pdsflx23.ne	regress_1	6/29/2005 12:34

Queue output without using test object Tags

It is possible now to kill any test testing BLK_FIFO_1 if it is found to be broken in this release of RTL

JOBID	USER	STAT	QUEUE	FROM_HOST	EXEC_HOST	JOB_NAME	SUBMIT_TIME
3621	mendonca	RUN	short	jupiter.a.eng.r.com	pdsflx23.ne	psax_status	6/29/2005 11:32
3851	shiv_runall	RUN	short	ganymede.eng.r.com	pdsflx23.ne	BLK_FIFO_1	6/29/2005 10:30
3852	shiv_runall	RUN	short	ganymede.eng.r.com	pdsflx23.ne	BLK_FIFO_1	6/29/2005 10:31
3853	shiv_runall	RUN	short	ganymede.eng.r.com	pdsflx23.ne	BLK_FIFO_2	6/29/2005 10:31
4001	robjose	RUN	short	saturn.eng.r.com	pdsflx23.ne	fifo_full_1	6/29/2005 12:31
4002	robjose	RUN	short	saturn.eng.r.com	pdsflx23.ne	fifo_full_2	6/29/2005 12:31
4040	ribbit	RUN	short	saturn.eng.r.com	pdsflx23.ne	frame_stat1	6/29/2005 12:34

Queue output using test object tags

Figure 8.5. Queue Management using a Tag Mechanism

8.5 Step 5: Test Case Identification

8.5.1 Structure of a Test Case

The next step is to identify the test cases that are required to test the design. A test case is one of the basic ingredients of any verification environment. A well written test case can be reused many times depending on the circumstances. Prior to identifying the test cases, a brief discussion on different types of test cases is presented.

There are a number of ways to write a test case. However, an analysis reveals that a test case has usually three parts to it. These parts are can be broadly classified into the following three parts:

1. Test setup section.
2. The test body section.
3. The test end (or cleanup) section.

Depending on the environment built for testing the design. The test setup section and cleanup sections are typically present either implicitly or explicitly in the test cases.

If the environment is built such that the tests modify or extend existing functions present in the environment, then the test setup and cleanup parts are deemed implicit. Many environments built using the HVL's typically fall into this category. Having implicit or explicit sections has no bearing on the effectiveness of the verification environment. Both in my opinion are deemed equally effective. They are merely different ways to achieving the same goal.

However, from a management perspective, there could be a difference. A verification engineer who is new to the environment has to typically go through study, and ramp up, before he/she is able to write tests when implicit setup and cleanup are used. In projects where the verification cycle for a module is very short or compressed for time, this may become a factor in determining whether additional engineers would be in a position to help the effort or not. The choice of the strategy also dictates whether all engineers know the verification environment infrastructure or whether a lesser number of people will be able to help maintain the environment.

A typical example of a test case with various sections is depicted in figure 8.6.

Test Setup section

This section is typically used to setup the environment to run the particular test. The section may comprise of reset code to reset the device, and possibly other programming information. It may also include test code to bring the device to a certain state so that a test can be done. In some cases, some memory

```
Void globals () {

    #define INIT_PACKETS 10
    #define LAST_PACKET_COUNT  20

    #define OFFSET_HEADER 6

    extern int CurrentPacketCOunt;

}
//----------------- Begin test setup section ---------------------------------------------
Void init () {

    tbSet.DumpEnable = FALSE;           //If you want the signal dumps
    tbSet.StartDumpPktNo = 0;         //Starting Pkt no for signal dump
    tbSet.StopDumpPktNo = 0;
    tbSet.AbortSimulation = FALSE;
    tbSet.EnableShortPkts = FALSE;

}
// ------------------- End of test setup section --------------------------------------
Void test () {

    init() ;

    // Programming the payload and the CRC byte of some of the packets
    if(PacketStream[MAC1_1].CRC==0)
    {
     SetCRCError(packet[3], 0, "00", FALSE, mac[1]);
     else
    {
     packet_no = packet_no * 100/4; // every 25 packets
     SetCRCError(packet[3], 0, "01", FALSE, mac[1]);

    }
......
...other test code
....
}
//---------- test cleanup section --------------------------------------------
Void cleanup() {
// all cleanup actions go here
for(int I = 0; I < MAC_COUNT; i++) {
Delete PacketStream[i];
}
```

Figure 8.6. Test Sections Example

required by the test as well as checkers, may be initialized and brought to operational mode in this section.

The test section

This code is portion of the actual test code that causes specific condition(s) to occur in the device. This code could be of many forms. A brief example of this code is given above.

The test section is the actual code. While the setup and end sections of a test may be common to all tests, this section is notably different from one test to another.

The test completion section

This section could be either in the test or in the environment. Typical actions taken in this phase of the test include checking the tests for error or releasing allocated memory or generating some statistics etc before the test ends.

8.5.2 Test Case Classifications

Tests are the essential ingredient in any methodology. In the section that follows, we present a taxonomy that classifies the tests into a few categories. This division of tests is based on the characteristics of the tests.

8.5.2.1 Directed Tests

The tests are usually the first tests to be written

These tests are usually procedural types of tests. They are used to address a specific scenario or condition in the device. These tests have the properties listed below. In many verification environments, this kind of tests is typically used to debug the RTL and the environment at the early stages. This is because the directed tests can be written to create a specific condition.

For example: Consider a device that uses an 8 bit microprocessor bus as shown in figure 8.7. This bus supports read and write transactions. Hence, a test to read and write from a specific memory location may be considered a directed test.

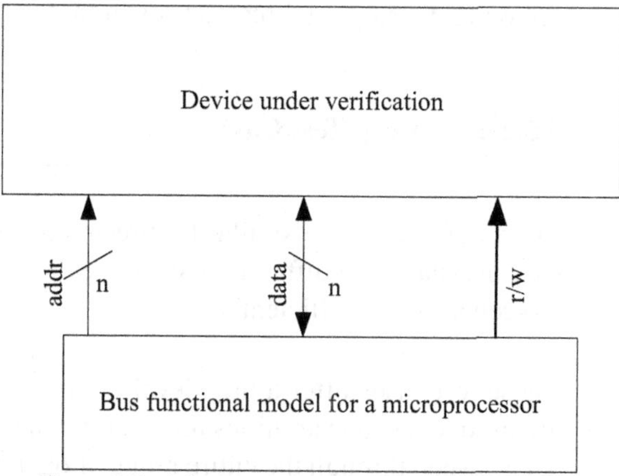

Figure 8.7. Microprocessor Bus Test Case Example

Directed tests are repeatable with the same results

The tests have one important quality: Consistency. Every time the test runs, they produce the same results.

They are present in almost all environments

In a random verification environment where the values of the inputs are determined randomly, the directed tests can be produced by merely constraining the random generator to produce a specific value every time.

They are still an important test method

Many designs that use a datapath typically use a set of directed tests to ensure that the tests are functional. Some designs do not benefit a great deal from a random methodology and use directed tests instead.

Moreover, in some designs, depending on the randomness of the test, it may be very difficult to create some specific scenarios to test the device. It is possible that the test may have to run for a long time to create the scenario. Under these circumstances, a directed test may be deployed effectively to test the device. (see the chapter on *Doing it right the first time*.)

In some cases they may be impractical

In case of a large SOC with many functional features, the list of functional features may run into the hundreds or thousands as the case may be. Using a directed test case for each scenario will quickly become a very daunting task. The task of maintaining them also becomes a big challenge very quickly.

Using a random generator or a tool that generates the tests instead of coding the tests by hand is frequently a good alternate option to developing directed tests in these circumstances.

8.5.2.2 Sweep Test Cases

Sweep test cases are similar to directed test cases. However, they are classified differently since they tend to cover a range of scenarios very efficiently.

These test cases offer a high level of stress in the design and are used to test on variations for a variety of clocking/protocol. They sweep through the entire range of operation by repeating a scenario across the design with a variation in time.

Why do these test cases even merit a different classification? This division presented by the author is deliberate. Having a separate test category for this test case type helps the test writer to think "out of the box" and otherwise identify test cases that the writer may not have thought of. The author has been able to find some rather interesting bugs quickly rather than wait to find bugs by generating the sequences randomly.

To further illustrate the idea of sweep test cases, consider a bus interface block which interfaces between two subsystems. This bus interface block interfaces between a internal system bus called LX bus and a system bus controller(may be a PCI type bus or some other similar bus for example!) This module is shown in the figure 8.8. When a request for date is made on the LX bus shown in the top half of the figure, the bus interface module translates the request and sends it to the system bus based on several parameters. It does so by raising the REQ line to the system bus controller.

The system controller gives back a GNT signal as shown in the figure. The GNT signal signifies that the bus is available for use. Such a protocol is shown in the figure 8.8.

In this example, it is specified that the response from the system bus could come anywhere from 2-20 clock cycles as per the specification of the device. Any response to the device less than 2 cycles and more than 31 cycles and less than 48 cycles is considered an error. Any response after 63 clock cycles is

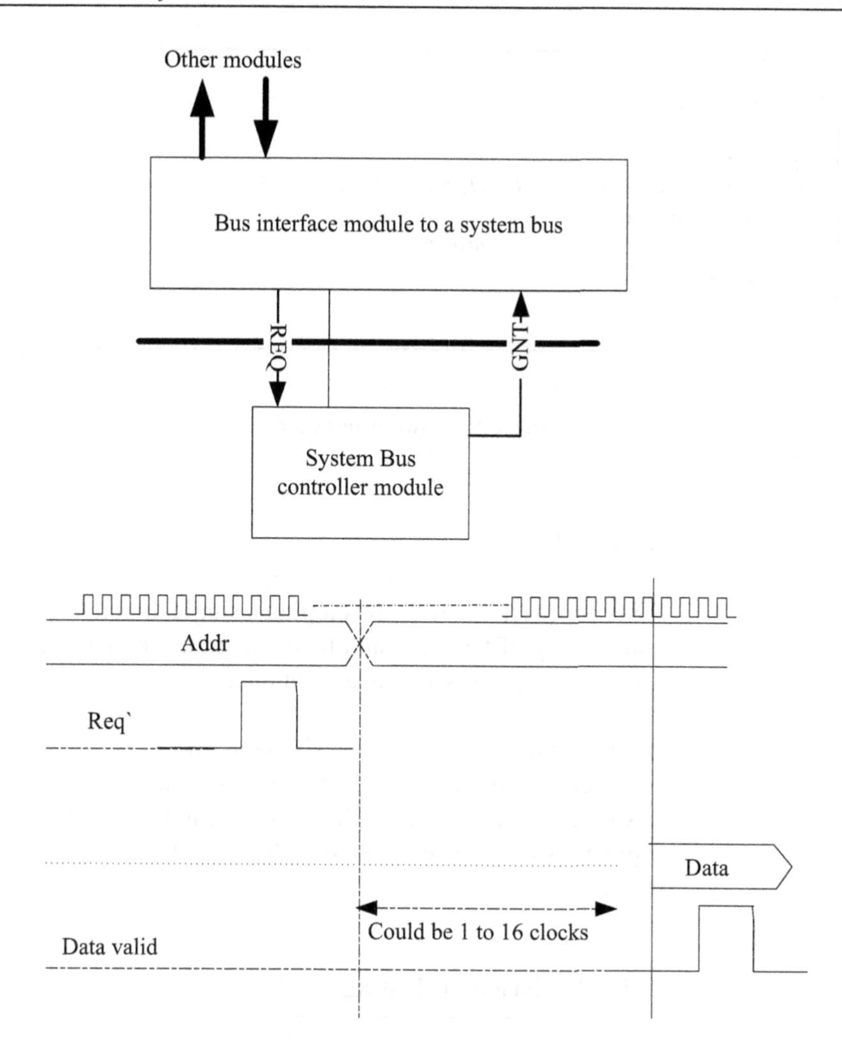

Figure 8.8. Sweep Cycle Test Cases

ignored and considered a timeout.

Looking at the module and the specification, one can write a few directed test cases to exercise the device. However, the full gamut of test cases becomes a challenging task. One approach is to set up some monitors and then use random responses to effectively test the design. This is also a valid approach.

A third alternative is to use some code that generates the following test sequence. The pseudo-code for this is illustrated in

```
For (delay in 1..70 cycles)

 Begin loop
   Do a transaction do_write(address1) from LX bus
   Wait for  delay clocks
   Send response from the system controller.
    Delay = delay + 1;
  end
```

Figure 8.9. Sweep Test Cases

figure 8.8.

This approach involves developing a single directed test to ensure sanity of test case and then extending the test case. All possible responses are covered at once.

In the author's opinion, a test similar to the above could be developed very easily after the completion of the development of a single directed test case. The incremental cost of test development is very low compared to the benefits this style of cases offers.

8.5.2.3 Negative Testing

Negative cases are those cases which exercise a design *outside* the given specification for the device. These cases are characterized by intending to create erroneous behaviors in the device and then observing the response of the device. An example is writing to a FIFO block when the FIFO is full. Another example is writing to a read only register.

Negative test cases are sometimes characterized by needing some sort of special handling. These test cases may cause some assertions in the environment to fail. In some cases, the fact that the design failed due to the erroneous inputs is actually a "passing" test case since it did produce the correct behavior from the device. However, in automated testing environments, these tests may prove to offer a challenge in making sure that the status is indeed reported correctly.

It is noted that every Bus Functional Model and Checker or Monitor will support such negative testing by default. The importance of not designing this into the environment cannot be overstressed by the author.

8.5.2.4 Random Test Cases

One of the main advantages of random testing is that it now becomes possible to uncover bugs faster than using directed tests. However, it must be realized that many a time, the random generator may not be in a position to generate specific random sequences without extensive intervention of the verification engineer. This may be due to several factors: The quality of the constraints placed on the random generator, the random generator itself and the nature of the environment and support for random test debug.

One important thing that must be noted during the use of random generators is that the output of the random generator must be continuously monitored to make sure that a large state space is indeed being generated by the random generator. This flow is shown in figure 2.10.

Directed random tests

This class of tests is typically found where the random generator is constrained to produce a subset of the possible values to test a specific area of the device.

Interactive random tests

The author has chosen to classify this class of tests separately. In many verification environments, it becomes necessary to test the interaction of any one feature with many other features, for example a SOC. A CPU with many different types of instructions is a good example. In this class of tests, there is a large state space. Functional coverage metrics and other metrics are usually used to determine whether the requisite state space has indeed been met.

Random interactions between different features is typically a well known method to rapidly uncover any issues of interaction between blocks.

8.5.3 The Creation of a Possible List of Test Cases

The steps in the previous sections were focused on creating the list of test objects. In this step, we create a list of test cases based on the functional list of objects.

In this section, we attempt to create a list of possible test cases by extending the list of test objects that were created in the earlier step. This list attempts to find out which of the following kinds of tests that were described earlier are actually applicable to a given test object. A summary of the different types of tests is provided below.

1. Directed Tests
2. Sweep Tests
3. Clock Crossing Tests - Tests that exercise designs with multiple clock domains.
4. Error Tests or random tests
5. Random Directed Tests
6. Random Interaction Tests

In order to create a effective test case list, the following questions are then asked *for each of the test objects on the list.*

1. What are the properties of this test object?
2. What range of values can be created for this test object?
3. What is classified as an error for this test object? Do any negative tests apply to this object?
4. Does this test object actually encompass multiple clock domains?
5. Is there some sort of a protocol associated with this object?
6. Are there interactions between other features or properties for this Object?

And so on. One may choose to customize the categories and classes of tests based on the relevance of the types of tests and environmental support available. Doing so helps the verification engineer identify specific scenarios that may be used to test the device.

In some cases, the test object may be of a bus type response. If this is indeed true, it may be possible to use the sweep method to attempt to test the test object exhaustively for some specific conditions.

If a clock crossing feature is present in the test object, it will become necessary to include tests to verify the design for these conditions as well.

Erroneous behavior is also determined for each test object. For example, writing to an read-only register. Typically doing something outside of the region of specification would classify the test as an negative test.

In some situations, it will become impossible to write a huge number of directed test cases. This is becoming true today. In such circumstances, tools like Specman[5] and Vera[6] help the verification engineer to choose a random or a coverage driven strategy to test the device. Hence we can create a plan as shown in the picture 8.10.

The grid shown figure 8.10 has the list of features on one axis and the types of test cases/monitors etc on the other axis.

Initially, it may be possible to just capture the intent of the design in a brief review. This intent may merely state which test categories are applicable for a particular test. In the figure 8.10, the intent may be captured by marking an 'x' in the appropriate column. Alternatively, the team may choose to describe the test by describing the scenarios as indicated.

[5] Specman is now a trademark of Cadence Design Systems.

[6] A description of Vera can be found on http://www.openvera.org Tools for this language are sold by some vendors including Synopsys.

1. Sweep the cache with port B write and Read for same address
2. Sweep Arbitration for different types of accesses

Block Name	Feature Object	Directed tests	Sweep Tests	Clock Crossing	Error	Random Directed	Random interactive	Test Object Tag
Cache	Cache arbitration for Port B read	x	x		x	x	x	Cache_2
Cache	Cache arbitration for Port B write	x	x			x	x	Cache_4
Cache	Cache arbitration for Port A read	x			x	x	x	Cache_1
Cache	Cache arbitration for Port A write	x				x	x	Cache_3
Cache	Cache flush algorithm		x		x			Cache_5
Cache	Cache controller states		x		x	x	x	Cache_6
Fifo	Functional soft reset	x			x			B1_RST_1
Fifo	Insert 0 pattern	x				x	x	B2_0_PAT_1
Fifo	Autoincrement feature		x	x		x		B1_AUTO_1
Fifo	Frame timing reference			x		x	x	B1_FRATIM_1
LLC	Different n bit positions for L1 and L2		x	x			x	B2_NBIT_1
LLC	Continuous output pointer increment					x	x	B2_OPINT_1
LLC	Payload crc insertion							B2_PAYL_1

Figure 8.10. Template for a Verification Plan

Some other alternatives that can be practiced in this step

Initially it might sound like a lot of work to actually come up with a plan in this manner. However, it is possible to undertake many shortcuts in this step. All the appropriate scenarios can be marked on this plan. The actual test descriptions and test intent can be saved for later and completed when the actual verification is done in a succeeding tests.

At this time, it must be recalled that all that has been accomplished so far is an analysis using a pen/paper or a computer spreadsheet or other means. The actual test definitions will be automatically extracted from this work process.

Do not attempt to optimize the test cases at this point yet!

Looking at the completed plan, It might be noted that the reader may begin to wonder that there is a large number of test cases, Some of them might be redundant and some of the tests aren't of high importance or unlikely to occur. A priority can be easily assigned at this stage or later. Optimizing the test cases is done in the following steps.

The intent of this process is to identify if there are any specific scenarios that need to be tested. The recommendation is that the identification of scenarios be done without any attention being paid to the actual work that this step may generate. The actual trimming of tests will be done in a subsequently in step 8.

Necessary and complete criteria are defined

Following such an approach allows the test writers to complete a necessary and complete criterion without being concerned about the actual work. If optimization were carried out earlier, the issue that crops up is that not all parameters are considered when tests are pruned or optimized.

Some of the features may be completely covered at other levels and some other features will be covered by monitors or checkers. When these are marked off in the next couple of steps, then the actual pruning work becomes very clear and very simple. A detailed example is presented in Step 8.

8.5.4 Partitioning of Tests between Block and Top Level

In the previous sections, we had built a list of features for the device without looking into the testability aspects of the features. As the features are analyzed, it becomes apparent that some of the features of the device are buried deep in the modules of the device. In some cases, the features of the module may have an effect on other modules.

The partition for testing the feature at the top level or at the block level is dependent on some factors. If the feature is buried deep in the module under test and does not affect other modules, it may be a good idea to test this feature in a module level. If there is interaction between other modules for this feature, it becomes apparent that the feature needs to be tested at the top level. *It must be noted that any features extensively tested only at the module level may indeed lose gate level coverage.*

For example, an arbitration signal to an external arbiter module will affect the behavior of the other module. However, a module specific item like a tap point coefficient in a digital filter may have some effect on performance, but the operation of the filter itself can be well tested at the module level. The behavior of the system with some specific filter settings could of course be tested at the top level. (This would obviously be a separate test object!)

Consequently, it now becomes essential to identify features and properties of the module that have an global effect on the device operation. These features would essentially need to be tested (in an order of priority of course!) at the top level of the device.

Looking at the table, we can add a column indicating to the user the various features at the top level. The tests and features can then be sorted as shown in the figure so that an report could be extracted

Block Name	Feature Object	Directed tests	Sweep Tests	Clock Crossing	Error	Random Directed	Random interactive	Checker	Monitor	Unique name	LEVEL of tests
Fifo	Functional soft reset	x			x			C1		B1_RST_1	chip
Fifo	Insert 0 pattern	x				x	x	C1	M2	B2_0_PAT_1	chip
Fifo	Autoincrement feature		x	x		x		C2	M1	B1_AUTO_1	module
Fifo	Frame timing reference			x		x	x	C2	M3	B1_FRATIM_1	chip
LLC	Different n bit positions for L1 and L2		x	x			x	C3	M2	B2_NBIT_1	module
LLC	Continuous output pointer increment					x	x	C4	M1	B2_OPINT_1	module
LLC	Payload crc insertion							C5	M1	B2_PAYL_1	chip
Cache	Cache arbitration for Port B read	x	x		x	x	x	Cache 2	m3	Cache_2	module
Cache	Cache arbitration for Port B write	x	x			x	x	Cache 2	m3	Cache_4	module
Cache	Cache arbitration for Port A read	x			x	x	x	Cache1	M3	Cache_1	module
Cache	Cache arbitration for Port A write	x				x	x	Cache1	m3	Cache_3	module
Cache	Cache flush algorithm		x		x				m3	Cache_5	chip
Cache	Cache controller states		x		x	x	x		m3	Cache_6	module

Figure 8.11. Partitioning of Tests at Block and Top Level

A review process will be able to identify if some specific features are missing from any particular list with ease!

Why is this review process helpful? One of the main benefits of approaching the test objects as described above is that it helps the verification team to very quickly identify the features that ought to be tested at the top level.

Many a time, the top level environment may enforce a coding requirement that can be taken into account when the module level tests are being written. Such an approach will minimize the effort of porting of the tests later on in the environment.

At the end of a project, it becomes a fairly trivial task to identify which features are tested where. This review is crucial. The review usually helps ensure that the features needing attention at specific levels did indeed get the attention that they deserved!

Since all the features needing to be tested at various levels have been identified, a review process can then be used to hierarchically identify if a feature tested at the block level ought to have been tested at the top level or vice versa. This review can be very easily facilitated by producing a few lists with the following information:

- A list of features that are tested only at the block level.
- A list of features that are tested only at the top level.
- A list of features that are tested at both the top level and the block level highlighted separately.
- A list of features that were missed out and need to be looked at to determine if they are redundant.

Any error from any of the lists usually stands out as a glaring omission on the combined list #3 above.

8.6 Step 6: The Definition of a Correctness Strategy

In the previous section, we identified the various possible test scenarios that were possible to test the various test objects in the device. In this step, we attempt to create the definition of a correctness strategy for the device that will be verified. The checkers that are described in this section may choose to implement monitors internally. However, for reasons that are described earlier, the checker is expected to perform only the function of checking and leave the function of observation to the monitors.

The figure 8.12 gives an indication of the various types of checkers and their positioning in the verification environment.

As can be seen from the figure 8.12, the data and cycle accurate checkers may be connected in parallel with the device. The protocol and interface checkers may be connected on specific interfaces of the device.

8.6.1 Data Checkers

Data checkers are typically used to verify the integrity of the data that is being transmitted through the device. These checkers are typically connected to the inputs and outputs of the device as shown in the figure. The datapath checker typically maintains a copy of the data that was passed into the module. It then examines the data that is made available on the output. The checker then ensures that the data was not corrupted or otherwise changed when the data passed through the module. In some cases, the checker may be able to predict the data which may be altered by the device under test presented at the outputs of the module under test.

Writing a data checker should be undertaken by taking into account whether the data checker may possibly be reused in environments other than the module level alone.

8.6.2 Protocol Checkers

The protocol checker is a checker that checks the protocol on the interface where it is connected. It usually flags an error if there is a problem with the protocol. A couple of examples for these checkers are the PCI bus or the IIC bus which is used by some devices to communicate with other devices on the system that they are designed into.

Some of these checkers are available from third party sources as verification IP. Given the high amount of integration, that most new SOC designs are experiencing today, outsourcing these checkers is becoming commonplace.

8.6.3 Interface Checkers

Interface checkers are checkers that will verify the interface of the device. These checkers usually are capable of identifying whether something has indeed gone awry on the module boundaries. These checkers may under some circumstances be protocol checkers if the modules exchange information via a well defined protocol that has been coded using the protocol checkers. In other cases, they may manifest themselves as assertions that verify the absence of illegal input to the module that is being tested.

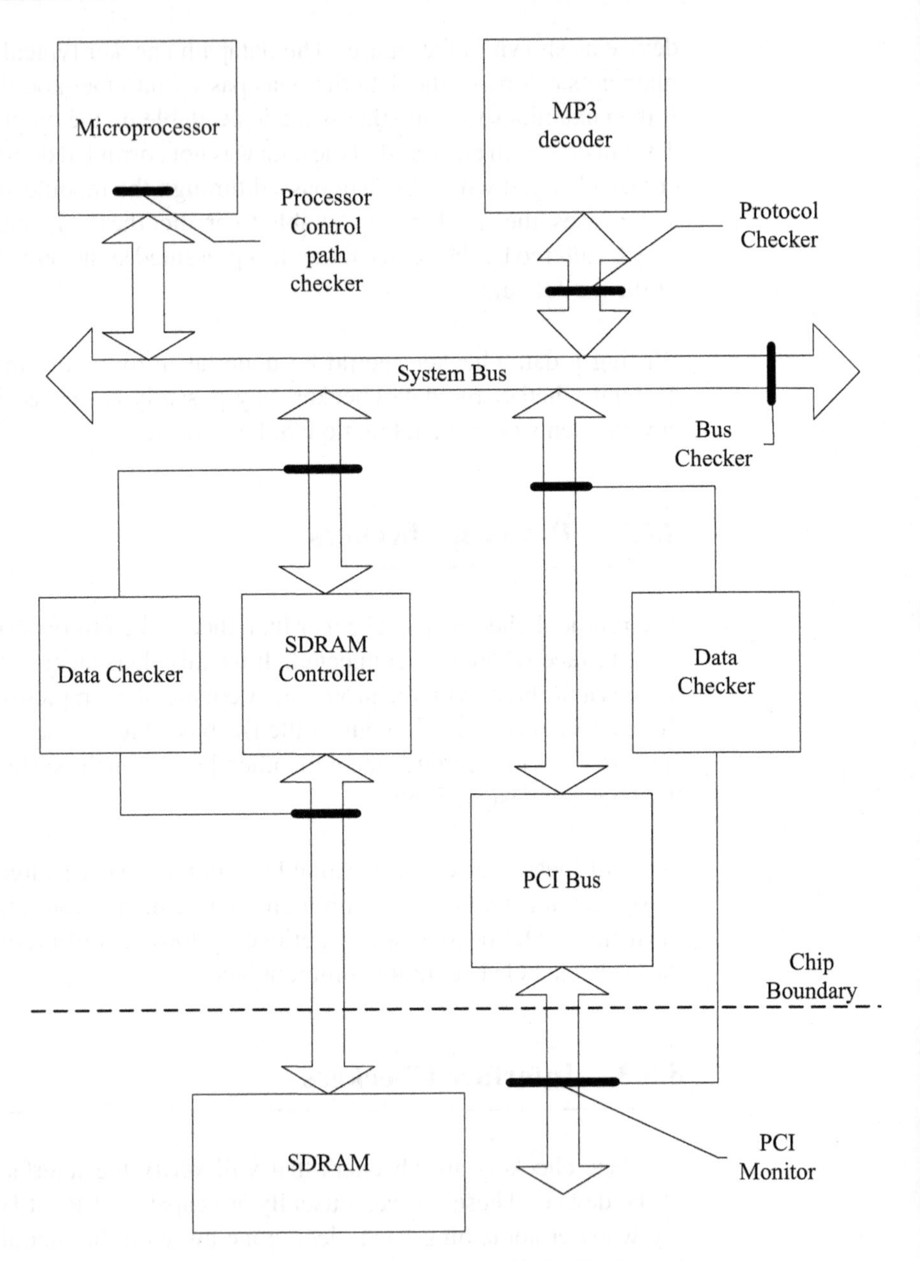

Figure 8.12. Checkers in a Verification Environment

Some of the interface checkers may also be placed on the outputs of the module to make sure that the module is indeed giving the right responses on the interface.

8.6.4 Cycle Accurate Checkers

Cycle accurate checkers are checkers that are used to check that the output or other monitored points of the device are correct on a cycle by cycle basis.

Cycle accurate checkers are usually difficult to write. These checkers may in some embodiments be simple by obtaining their output from a golden model which is a accurate representation of the design. In other embodiments, they may have extensive checking functionality built into the checker itself. Since the design is usually subject to significant change early on in the design cycle, maintenance of the checker may be difficult. However, depending on the circumstances, the checker will be a very good way to identify issues when the design does not meet or match the specification.

8.6.5 Using Monitors

Monitors are used to observe the design and print warning messages if needed into a log file. They serve as a important tool in debugging the design completely since they capture information about the events in the design into a log file and allow a detailed analysis to be completed after the simulation has been completed.

Monitors can be created and instantiated into the verification environment using a variety of methods. Some engineers have chosen to use HVL's like Vera or "e" which offer ease of use. On the other hand, some of my colleagues have chosen to implement the monitors using various testbench constructs available in the HDL itself. Various considerations on writing monitors are presented in the previous chapter.

8.6.6 Using Assertions and Formal Verification in the Methodology

The previous sections described the derivation of test objects and the process of making sure that the test object list is a complete and accurate one. In this section, the possibility of using assertion based methodologies is explored.

Assertion based techniques have many advantages. These are:

Assertions in simulation

Using assertions in a simulation environment helps the user to leverage the infrastructure that is present in the existing simulation environment. During the simulation, assertions can be used to monitor activities inside the design and on the interfaces. When a failure occurs because of an assertion, the problem can be flagged immediately. In many cases, this timely warning allows the testbench to detect an error and take appropriate action before waiting for the results after post-processing etc. The assertions help with "bug triage" and allow the cause of any violation to be easily identified.

Coverage

Since the assertions are now embedded in the design, the assertions can also serve as coverage monitors. They can be configured to collect statistics and coverage information for various parts of the design. Once the data is available, it can be processed to ensure that design has been simulated for all possible cases.

Dyamic formal simulation

A third benefit from using an assertion is that it now becomes possible to use a formal verification tool to aid with finding bugs. Formal analysis can also be used to determine if the design exhibits some interesting corner-case behaviors. If a violation is discovered, formal analysis will provide a counterexample to the assertion. Since stimulus from simulation is leveraged to find the violation, the counterexample is something that can be found as an extension of the existing stimulus and is a whole lot easier to debug.

In order to be able to effectively deploy assertions in the methodology described herein, the author proposes some simple guidelines that may be modified to the needs of the user.

Map each test object to a collection of assertions

Each test object on the test object list can now be mapped to one or more assertions in the assertion list. At this juncture, what is required is that the test object must be completely specified in terms of assertions. If there are any temporal dependencies for the property, i.e. the test object has some time related properties as well (can be easily seen with sweep test cases etc) then it is apparent that the assertions will not be merely static but also temporal as well. This is seen in the figure 8.13. The

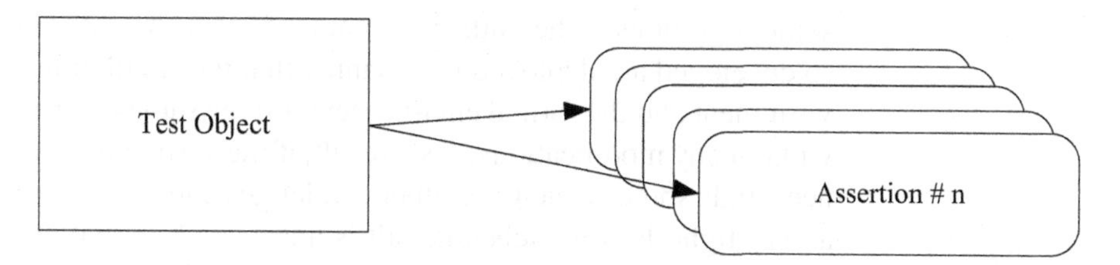

Figure 8.13. Mapping Assertions to Features

key side effect of this organization is that it becomes possible to determine if the list of assertions for each test object is now a *necessary and complete set of assertions.* Any review of the assertion list can now focus on answering the question of completeness of mapping.

A coding style that tells you what the assertion does and is mapped to is a must

The use of a coding style where it is evident from the assertion instance the feature of the design is a must. The author would like to propose that the ***test object tag*** be a part of the instance name. This will allow the verification engineer to quickly share information with the design engineer specifics of the failing feature and thereby get the debug cycle done more quickly. In the figure 8.14 , it is apparent that the test object cache1 has several assertions Cache1_assertA_port1, Cache1_assertA_port_wr_check etc. As a result of this organization, the verification engineer can now focus on the following and other important aspects:

- Did the intent required for this particular test object get defined?
- Are the assertions correct?
- Are the assertions enough and complete?
- Are the assertions in line with the test object? Ie. Are there temporal assertions if there is some time dependency?
- Is it possible to prove the formal verification with a tool?
- Can the assertions be reused as monitors?

Keep in mind that any formal methodology may give a indeterminate response

Formal verification techniques do offer the user complete proof of the assertion being made. On the other hand, the formal verification tool may suffer from some capacity constraints depending on how the tool is implemented. As a result, the verification engineer may find that it is very challenging to prove some assertions. The author recommends that the assertions so developed are deployed in a manner that the simulation environment and the formal environment use the same assertions without any modifications. As a result, if the verification engineer finds some capacity or other challenges and is faced with a very tight delivery schedule, all is not lost. A partial proof may be possible supplemented by simulation. This recommendation is made based on the observation that some verification approaches use two different approaches which have no possibility of overlap may actually cost the verification engineer a great deal of time and effort, not to mention stress!

In the example given it may be possible that the assertion *Cache1_portA_read_no_asser* could not be proven using a formal verification tool. The verification team then has the alternative of either decomposing the assertion into several smaller ones and attempting to prove them or using simulation as a backup option to verify the test object if there are some other factors at work like schedule, licenses etc. In spite of these challenges, Formal verification has found many bugs which have been deemed "hard to find" and instances of this can be found in literature. The modern verification efforts would be hard pressed to achieve closure of verification if formal verification is not part of the overall strategy.

It is well known that formal techniques have some capacity limitations. These limitations are being actively worked on in the industry. Until a solution is found, the default has been the simulation environment. For test objects where the formal techniques are unable to close the gap, the simulation method can be used. This enables that there is no single feature that is unproven because of the two different approaches.

Many modern testbenches [8] actually use dynamic simulation and formal verification to attempt to verify the design.

8.7 Step 7: The test strategy

The previous step identified various checkers and monitors and helped define a correctness strategy that helps to check the device under test. In this step we attempt to identify the tests that need to be developed to verify the device. The tests by design should be able to trigger various events that are checked

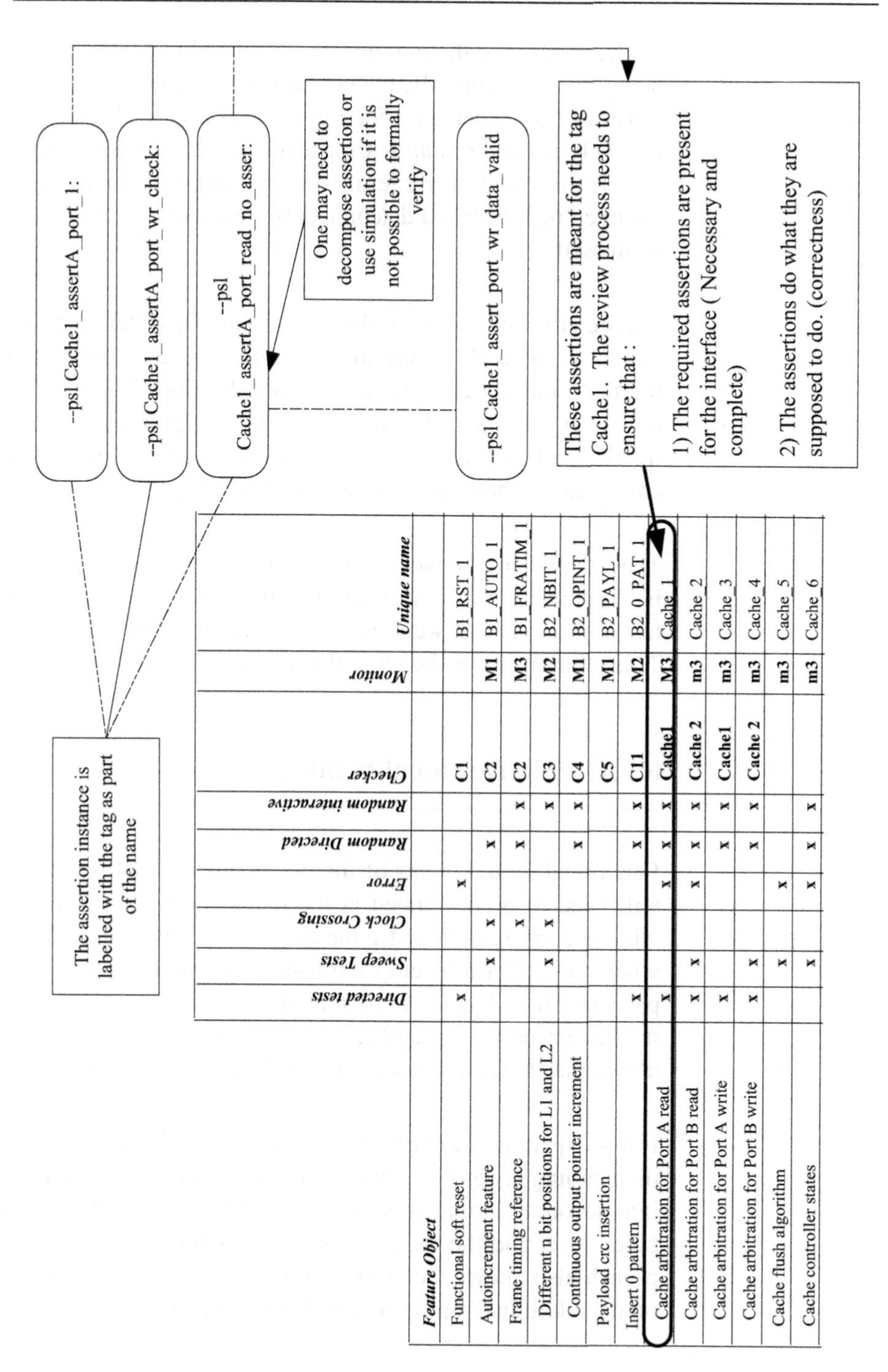

Feature Object	*Directed tests*	*Sweep Tests*	*Clock Crossing*	*Error*	*Random Directed*	*Random interactive*	*Checker*	*Monitor*	*Unique name*
Functional soft reset	x			x			C1		B1_RST_1
Autoincrement feature		x	x		x		C2	M1	B1_AUTO_1
Frame timing reference			x		x	x	C2	M3	B1_FRATIM_1
Different n bit positions for L1 and L2		x	x			x	C3	M2	B2_NBIT_1
Continuous output pointer increment					x	x	C4	M1	B2_OPINT_1
Payload crc insertion							C5	M1	B2_PAYL_1
Insert 0 pattern	x				x	x	C11	M2	B2_0_PAT_1
Cache arbitration for Port A read	x			x	x	x	Cache1	M3	Cache_1
Cache arbitration for Port B read	x	x		x	x	x	Cache 2	m3	Cache_2
Cache arbitration for Port A write	x				x	x	Cache1	m3	Cache_3
Cache arbitration for Port B write	x	x			x	x	Cache 2	m3	Cache_4
Cache flush algorithm		x		x				m3	Cache_5
Cache controller states		x		x	x	x		m3	Cache_6

Figure 8.14. Implementing Assertions in the 10 Step methodology

by the checker or the monitor that is related to the tests object being tested. Indirectly, the checker and monitor can be used to verify each other. I.e. For a given tests object, the tests that is developed and run *must* trigger the related checker or monitor. If this event of triggering does not occur it can be safely assumed that there is a disconnect somewhere that needs to be addressed.

This relationship between the test, monitor and the checker is crucial to the architecture of the verification environment. If the tests and the checker/monitor are developed by different people as is usually the case; the tests and the monitor effectively check one another. This gives an additional amount of confidence in both test and checkers/monitor code.

The relationship between the tests and the checkers/monitors is maintained in this verification approach is by means of the test object tags that were picked for each test object in step 3. This relationship is shown in the figure 8.15.

8.7.1 Hierarchical Strategy

Functional requirements of the device dictate that some tests will need to be developed at the block level and some tests will need to be developed at the chip level. This choice of level where the tests are developed is usually dictated by whether the particular test object is something that is buried deep down in a module or has a global effect. An interface between two modules at the chip level must obviously be tested at the chip level.

If it is apparent that a test object must be tested at the top level, then coding styles at the block and top level need to be taken into account. Doing so will make the transition between the block level and the top level proceed smoothly. Such an approach also allows the verification engineer to focus his time on the additional top level tests that need to be written.

Common instances of test objects that may span module boundaries are bus protocols and signals crossing module boundaries at the chip level as well as chip level reset etc.

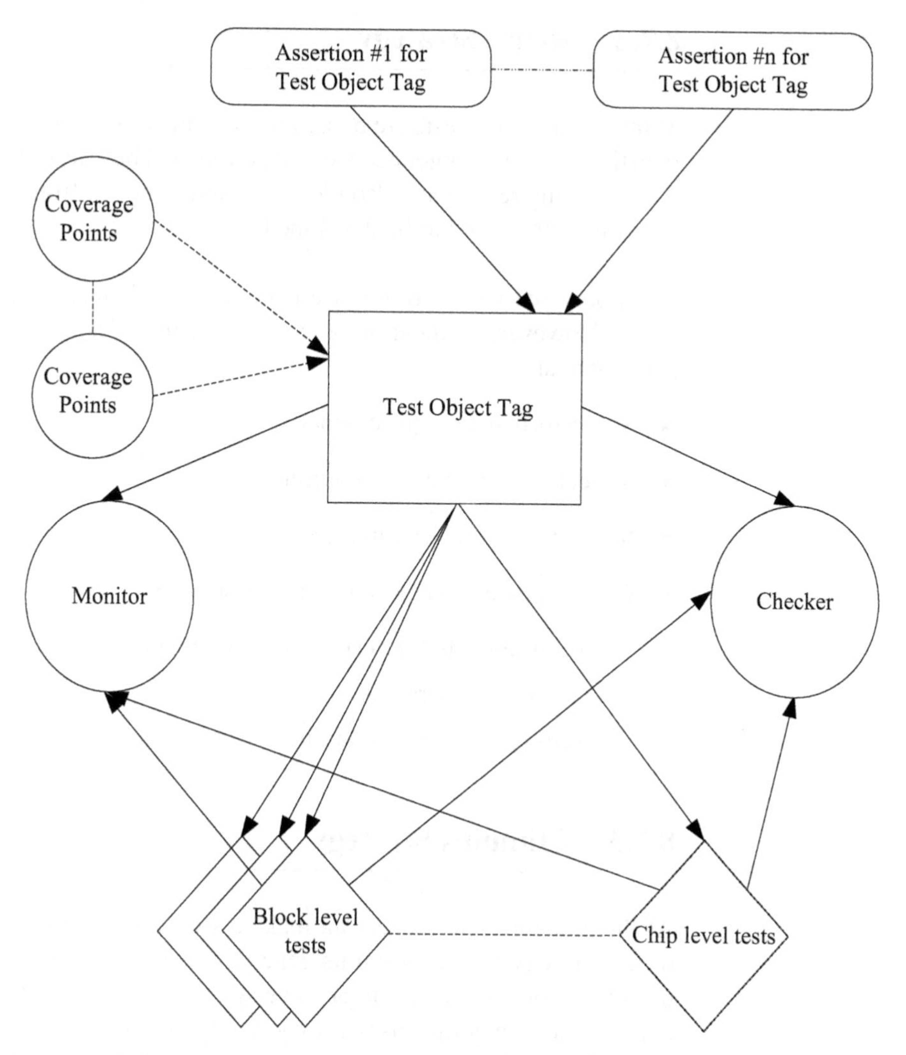

Figure 8.15. Relationship between the Tests and the Checkers/Monitors

A review of the test object list now reveals the objects that will need to be tested at the top level. These objects can then be identified in the test object list. Later on in this step, the tests are then linked to test objects. Since the objects are now required to be tested at the chip level (and possibly at the block level as well) a set of coding requirements is imposed on the tests by the environment automatically.

8.7.2 Reuse Strategy

Many a time, if the tests are being reused from another project, then these tests no longer need to be developed. These test cases that are being reused are also identified and removed from the list of tests that need to be developed.

One notes that verification reuse is not as prevalent as design reuse. However, verification reuse is easier than design reuse [9]. There are:

- No performance requirements.
- No technology-specific requirements.
- No synthesis tool requirements.
- What is however required to be defined are:
  - Requirements for verification components
  - Functional correctness.
  - Reusability in verification environment.

8.7.3 Stimulus Strategy

At this juncture, the reader can make a well informed choice of the strategy for the stimulus generation. Many a time, the use of random generators to achieve state coverage quickly is considered[7]. In some cases it may be the perfect solution to the problem. Usually a combination of tests helps complete the plan.

Control based devices (an example is a microprocessor) are typically complex devices which may benefit from a random environment. Verification of these devices usually incorporate random testing, formal testing as well as other methodologies to help address the large state space problem.

Datapath based devices usually are a little different than control based devices. In the case of datapath based devices, most of the data processing elements and functions operate on the data

[7] See the section on Doing it right the first time.

flowing through the device. In these devices, the operations performed on the data are important.

When verifying a datapath, one suggestion that is offered is to keep the environment as efficient as possible. The focus must be kept on being able to send large quantities of streams through the device. Hence, the overhead from the testbench and other components must be kept to a minimum. The penalty for ignoring this consideration is that the simulations and debug cycles typically tend to be long.

8.7.4 Test Case Strategy

The test objects in the test object list may be tested using a checker or monitor or a test case or an assertion based methodology.

It must be however noted that the assertion monitors must be placed on interfaces, corner case implementations and at specific points in the design, for which complete function validation is required (possible only by using formal verification). The location, quality and type of assertion monitor being used in a simulation environment will play an important role in the debug effort required after the detection of a bug.

- When using a simulation methodology along with a group of tests, results of the simulation can be used to measure progress.
- When using assertions, the assertions can easily be used to determine coverage using several metrics that have been discussed earlier.

The relationship between the assertions and the checkers/monitors is evident since the assertions are all related to the tests, monitors and checkers by means of the test object tag.

8.7.5 Identifying Test cases for Maximum Yield & Coverage

Various types of test cases and their applications were discussed in the earlier portions of this chapter. The test object list that has been developed so far contains a exhaustive list of possible scenarios that may be tested for the device. The earlier discussion had specifically avoided identifying test cases at that stage of development of the test object list.

At this juncture, all the information that is required to develop the test cases exists in the test object list. We have been able to identify test objects that are reused from other projects as well as test objects that need to be exercised at the chip level of the device. Specific assertions that can be used to test the test object have also been identified.

The test cases that give the maximum yield and Coverage can be picked right off the graph

Various types of test cases and their applications were discussed in the earlier portions of this chapter. The test object list that has been developed so far contains a exhaustive list of possible scenarios that may be tested for the device. The earlier discussion had specifically avoided identifying test cases at that stage of development of the test object list.

In addition, all the specific features that need testing at the top level have been identified. The testing and components to test the device at the block level or at the top level are also identified. This leads to some rather interesting derivations and conclusions as the reader will observe in the discussion that follows. *If the test object list is sorted by ROW, then it is now possible to identify the test scenarios that would be most effective in testing out the test object.* This is shown in the figure 8.16.

While the decision to choose a particular style of test cases is completely up to the verification engineer who is developing tests for the particular test object, all that needs to be ensured is that the test cases that are developed have taken into account the various features embodied in this test object (Range of values needing a sweep test, cross of clock domain boundaries etc, error handling etc).

All the test cases that are developed for the test object need to be in accordance with the plan that is presented above. In a similar vein, it is now possible to find the most efficient test case that tests a *group* of test objects. This is done by *creating a test that picks the most test objects sorted by a certain column*

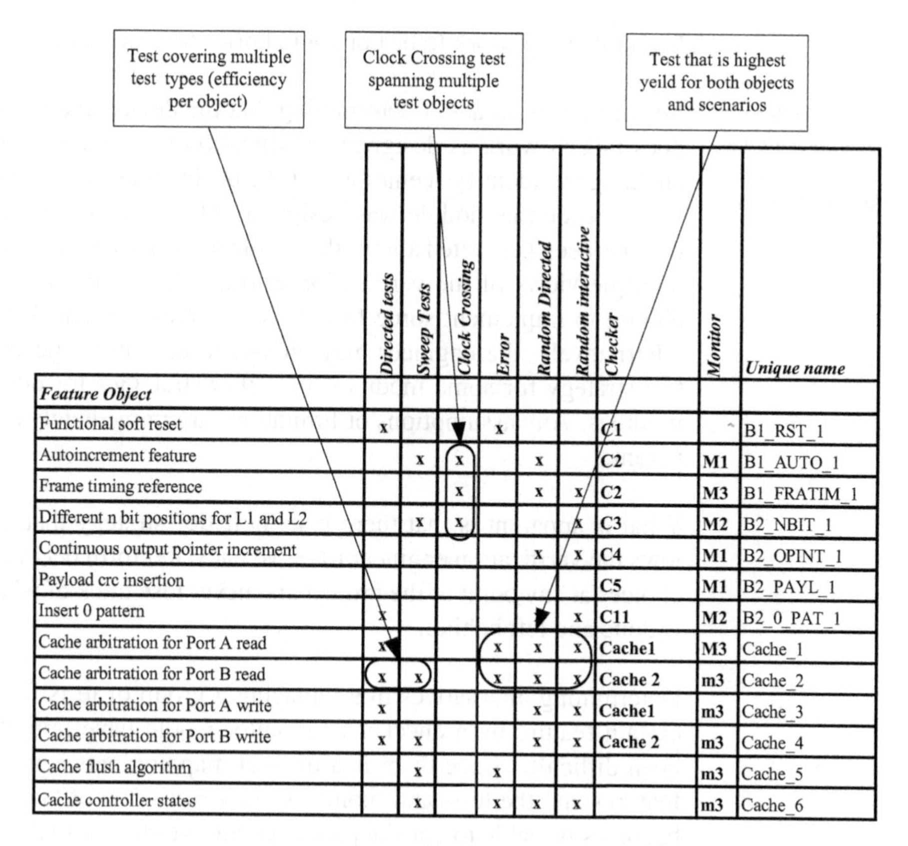

Feature Object	*Directed tests*	*Sweep Tests*	*Clock Crossing*	*Error*	*Random Directed*	*Random interactive*	*Checker*	*Monitor*	*Unique name*
Functional soft reset	x			x			C1		B1_RST_1
Autoincrement feature		x	x		x		C2	M1	B1_AUTO_1
Frame timing reference			x		x	x	C2	M3	B1_FRATIM_1
Different n bit positions for L1 and L2		x	x			x	C3	M2	B2_NBIT_1
Continuous output pointer increment					x	x	C4	M1	B2_OPINT_1
Payload crc insertion							C5	M1	B2_PAYL_1
Insert 0 pattern	x				x	x	C11	M2	B2_0_PAT_1
Cache arbitration for Port A read	x			x	x	x	Cache1	M3	Cache_1
Cache arbitration for Port B read	x	x		x	x	x	Cache 2	m3	Cache_2
Cache arbitration for Port A write	x				x	x	Cache1	m3	Cache_3
Cache arbitration for Port B write	x	x			x	x	Cache 2	m3	Cache_4
Cache flush algorithm		x		x				m3	Cache_5
Cache controller states		x		x	x	x		m3	Cache_6

Figure 8.16. Finding the Optimal Tests to be Developed

as shown in the figure 8.16.

The most valuable tests are the ones that cover both the rows and the columns

However, the test case that basically needs to be in a smoke or other regression is one that covers the *most from a row and column basis* Such a test case would be one that encompasses multiple rows and multiple columns. This is illustrated in the figure 8.16. While there might be better groupings and optimizations available more than what is shown in the figure, the grouping could otherwise be considered akin the Karnaugh map principle which is widely discussed in digital design books. It is only a symbolic representation of what is indeed possible.

Elimination of tests that are redundant is also possible

Some of the test scenarios may be covered by other test types in the verification plan. These tests can now be eliminated from the test development list. Any reused tests can be eliminated from the list of work to be completed prior to a tape-out

It is possible to generate multiple views and analyze them to cut down verification effort

Verification is all about determining that the device under test does indeed work as designed. Verification is also about the challenge to identify scenarios that the designer could not envisage when the module was designed. The verification plan that has been generated allows the verification engineer to have multiple views of the verification effort. The engineer may choose to implement some features as monitors or checkers. Alternatively, the engineer may choose to use some specific test strategy for some modules and other strategies for other modules. Any assumptions or limitations are immediately apparent.

What is apparent is that there is a "holistic" strategy that allows the verification engineer to move back and forth between choices at any point of the project and never lose track of what is done and left behind.

Determining alternatives like replacing a group of tests for a certain feature by a checker or monitor would have initially been difficult. Since there is a distinct mapping between the features and the tests and monitors via a collection of tags, it becomes possible to consider some changes during actual implementation of the tests themselves.

Change management is now a trivial task

As had been mentioned earlier, change is probably one of the more constant things in this world we live in. Frequently, due to a variety of reasons the feature list of the device may change somewhat. This may be due to engineering reasons or market forces. Since we now have a detailed list that maps to the test cases, checkers and monitors, all we now have to do is to identify which test objects indeed have changed. A corresponding list of tests, checkers and monitors to be worked on becomes apparent!

The important thing to be noted however is that the management of the entire test database can be done easily using automation.

At the completion of this step, the planning and identification process of the verification plan is complete. What now needs to be done is to create the appropriate scenarios and ensure that the design is tested.

8.8 Step 8: Testing the design.

8.8.1 Component identification

Identify the testbench components based on the items in step 7

The previous step helped us identify the various tests, checkers and monitors in the verification environment. Now, the hierarchical and reuse strategies are also well defined.

There is also all the information regarding the various features that needed to be supported by the device at the top or block levels.

Given the structure of our verification plan, it is now possible to sort the list by Module. The appropriate checkers and monitors that are instantiated in the design and testbench are immediately obvious.

There is an explicit specification in this representation for the checker

As can be seen from the figure 8.17, checker C2 must be able to check all aspects of The FIFO module's Auto-increment feature and the frame timing reference. It must operate across multiple clock domains and be able to handle inputs both *inside and outside* the specification of the above features. It must also be able to be used at the module level and at the chip level. The checker must be robust enough to handle random data as well.

There is an explicit specification for the monitor in this representation.

The second example that is presented is for monitor M3. This monitor operates with the cache module and must be able to handle a variety of inputs both inside and outside the specification for arbitration of ports A and B. It must be able to understand the cache flush algorithm and the cache controller states. The monitor may be instantiated at the chip level and at the module level.

In a similar manner, the specification for each checker and monitor may be derived. The tags that are common between the monitor and checker now help keep a balance.

Specification of Checker C2

Block Name	*Feature Object*	*Directed tests*	*Sweep Tests*	*Clock Crossing*	*Error*	*Random Directed*	*Random interactive*	*Checker*	*Monitor*	*Level*	*Unique name*
Fifo	Functional soft reset	x			x			**C1**		**Top**	B1_RST_1
Fifo	Insert 0 pattern	x				x	x	**C1**	**M2**	**Module**	B2_0_PAT_1
Fifo	Autoincrement feature		x	x		x		**C2**	**M1**	**Module**	B1_AUTO_1
Fifo	Frame timing reference			x		x	x	**C2**	**M3**	**Top**	B1_FRATIM_1
LLC	Different n bit positions for L1 and L2		x	x			x		**M2**	**Module**	B2_NBIT_1
LLC	Continuous output pointer increment					x	x		**M1**	**Module**	B2_OPINT_1
LLC	Payload crc insertion								**M1**	**Top**	B2_PAYL_1
Cache	Cache arbitration for Port B read	x	x		x	x	x	**Cache 2**	**M3**	**Top**	Cache_2
Cache	Cache arbitration for Port B write	x	x			x	x	**Cache 2**	**M3**	**Top**	Cache_4
Cache	Cache arbitration for Port A read	x			x	x	x	**Cache1**	**M3**	**Top**	Cache_1
Cache	Cache arbitration for Port A write	x				x	x	**Cache1**	**M3**	**Top**	Cache_3
Cache	Cache flush algorithm		x		x			**Cache 3**	**M3**	**Module**	Cache_5
Cache	Cache controller states		x		x	x	x	**Cache 3**	**M3**	**Module**	Cache_6

Specification of Monitor M3

Figure 8.17. Specification of Checkers and Monitors

Coding styles for the monitor and checker are apparent.

If there are any special considerations for the coding styles of the monitor and checker for either the top level or the module level, then these are immediately brought to the surface.

Any restrictions are also apparent.

Any architectural restrictions can be resolved and decisions can be taken by the team before the implementation begins.

There is an explicit definition of the BFM for a module in this representation

Explict definition is present in this representation. If there are any special considerations for the coding styles of the Bus functional model for the module level or the top level, then these are immediately brought to the surface as well.

In addition to these aspects, there is also the need to develop some scripts to enable the handling of regression, compilation and other housekeeping scripts. There is also the need to ensure that the bug tracking and other compute infrastructure essential to verification is now in place.

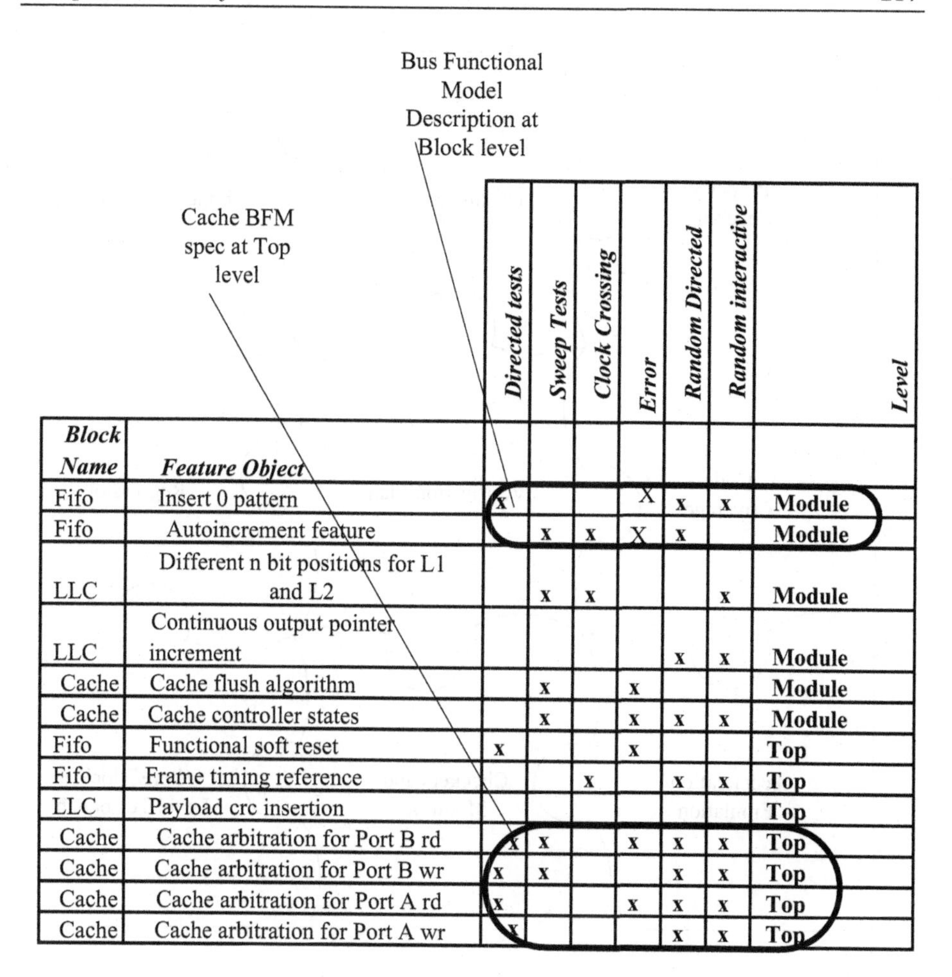

Block Name	Feature Object	Directed tests	Sweep Tests	Clock Crossing	Error	Random Directed	Random interactive	Level
Fifo	Insert 0 pattern	x			X	x	x	Module
Fifo	Autoincrement feature		x	x	X	x		Module
LLC	Different n bit positions for L1 and L2		x	x			x	Module
LLC	Continuous output pointer increment					x	x	Module
Cache	Cache flush algorithm		x		x			Module
Cache	Cache controller states		x		x	x	x	Module
Fifo	Functional soft reset	x			x			Top
Fifo	Frame timing reference			x		x	x	Top
LLC	Payload crc insertion							Top
Cache	Cache arbitration for Port B rd	X	x		x	x	x	Top
Cache	Cache arbitration for Port B wr	x	x			x	x	Top
Cache	Cache arbitration for Port A rd	x			x	x	x	Top
Cache	Cache arbitration for Port A wr	X				x	x	Top

Figure 8.18. Bus Functional Model Specs

Since the definition of various items is complete, it now becomes possible to ensure a complete view of the task that has to be done is assembled.

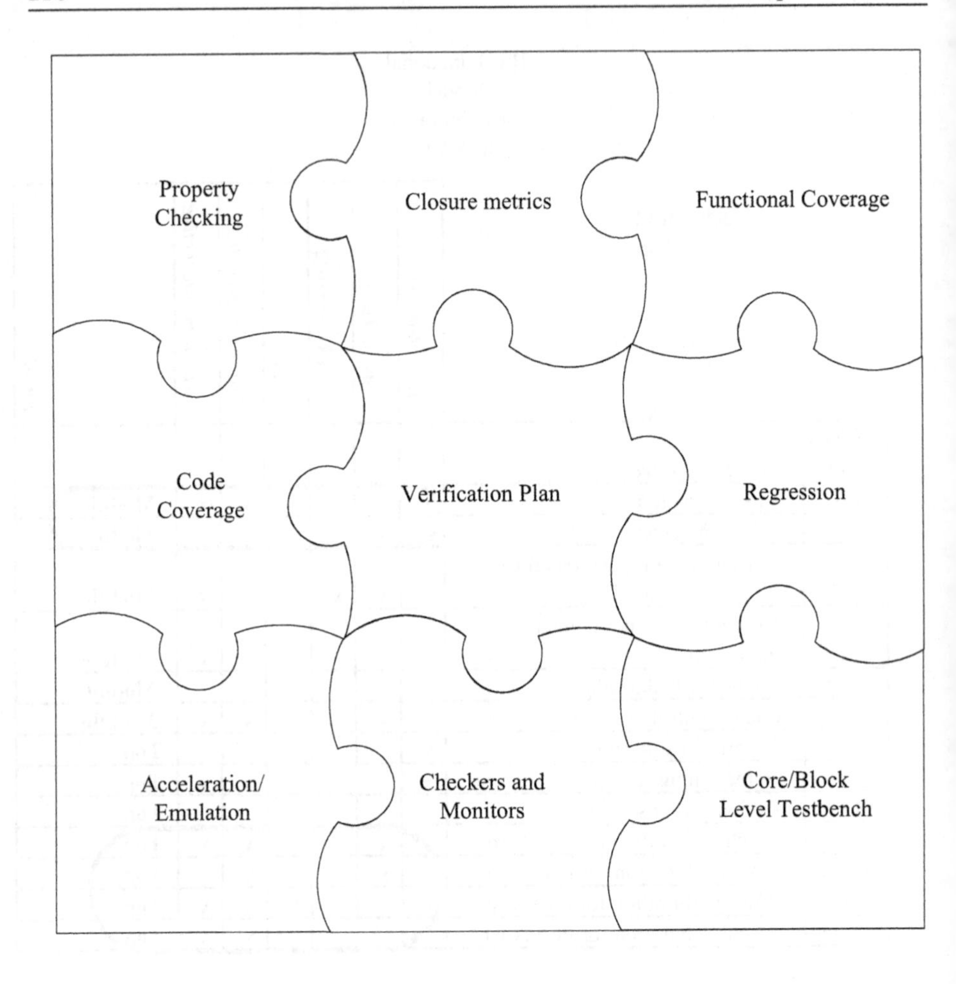

Figure 8.19. Various Aspects of Verification

We now know exactly what we have to do. What is required and what is necessary is now completely apparent from the verification plan

In the previous step, the test cases that need to be developed had been identified on the list. As can be observed, the test object list has now graduated to becoming a complete verification plan. This plan is complete from a testing point of view since it is able to now reveal the *necessary and sufficient* tests that are now required to test the device. The plan that is marked above now has a complete picture of what is necessary to test the device under test. All possible scenarios are mapped into one or other cells on the verification plan that has been created.

At the very minimum, this plan should now be reviewed for completeness and accuracy. The main intent of the review is to focus on the test plan and understanding of the test objects which has led to the test plan.

Completing a review prior to creating the test cases ensures that any assumptions in the plan are addressed. This essential step ensures that everyone in the verification team is now on the same page as far as the development is concerned.

We have been able to get a very good handle on the work to be done

We have also been able to optimize the pre silicon verification plan and the post silicon verification plans. In addition, we have been able to optimize the tests at the block level and at the top level.

The verification schedule is now apparent since we now know what we have to do

Now that we have identified the particular *types* of tests cases and the coding styles that are associated with each one, we have a rough handle on how long the verification effort will take. It is now possible to extract a summary from the verification plan that has been implemented. This summary can be easily generated to reveal what tests need to be written at the top level and module levels. A list of checkers and monitors are also available as part of the summary. Using some metrics on engineer productivity, we can then estimate easily what is involved in completing the verification plan. This is embodied in the calculation.

Any top level tests that need to be ported from the module level will probably impose a certain coding style or restrictions on what can be accomplished in a certain test. These are also apparent from the summary.

$$Total\ Days = \frac{Number\ of\ objects}{rate\ of\ coverage\ of\ objects} \tag{8.1}$$

It is now easily possible to divide the work in a fair manner among all the verification engineers in the team. Some of the verification team can attempt to complete the environment development while others focus on the tests or other elements that need development.

The progress of the verification effort can be be measured using available metrics

Many standard metrics like the bug find rate, the test object completion rates and the bug close rates now play an important role in determining the state of progress of the design. Since a list is now available, it becomes possible to take a educated guess at estimating how long verification is going to take. This estimate is something that in the author's opinion opinion will dictate tape-out schedules.

Any schedule is probably closer to the truth than a guess-estimate

During some earlier projects, typically, RTL design started as an activity followed by the physical layout design activity. The verification process was usually a "fuzzy" process that happened somewhere in between! Added to this was the fact that verification was frequently over cost and over budget anyways making estimation a challenge.

Many managers typically wind up giving a verification engineer a document and ask the engineer to come up with a schedule and a test plan for testing the device. Depending on the engineers past experience, this became a "shot in the dark" or a "shooting from the hip" kind of experience. The above process can be followed for even block level modules. The list that is generated now allows for both optimistic and pessimistic predictions on how long the effort will actually take.

Honestly, although the process may seem a little long winded initially, many parameters come very quickly under control. The verification plan is typically made available as a first draft shortly after the functional specification is released. Many a time, these efforts may go in parallel in many organizations. Keeping the test object list provides the engineers the necessary insulation against changes in the functional specification and having to rework everything.

Depending on the engineering skill sets and personal preferences, one can easily identify a week-to-week plan which they will follow. If something in the project schedule looks unrealistic (This happens a lot somehow, the reasons are unknown!). Immediate feedback can be given and appropriate risk mitigation is possible. In addition, there are instances where the author has used data from previous projects to request architects to leave pre-verified IP intact based on the cost of verification.

8.8.2 Getting the Job Done. Execution of the test plan

The previous sections concentrated on development of the plan and identifying the various components that need to be developed and verified. Now that most of the elements of the verification environment and test cases have been identified, the task of execution now begins. During this activity, various considerations from the *Cutting the ties that bind* chapter and the learnings from the case studies in *Doing it right the first time* chapters may be taken into account before the environment and test cases are constructed.

The beginning stage of this process is to create an estimation of the time involved in creating the tests and the environment. Since the previous steps had broken the entire complex task into a number of pieces, this task can be taken on and accomplished with ease as mentioned earlier. Any cross dependencies can also be worked out before execution of the plan. The test cases and the environment development work can then be divided amongst various team members and completed as per a schedule.

The process begins with the design engineer beginning development work on the RTL. In parallel, the verification environment, necessary scripts, and infrastructure are created. Based on the frequency of RTL releases, the tests are then run against the RTL after integrating the RTL with the testbench. Bugs are filed on the design when the design does not meet the specifications. (Bugs on the environments or tests as well!)

One of the recommendations of the author is to have a biweekly correlation of the tests and the state of the RTL in order to ensure that the test development is on track. This correlation described in the review section could be an informal one where issues of what worked and what didn't as well as the progress to date can be discussed along with any pressing issues that impede the verification effort. This way, any verification issue that arises will be addressed as soon as possible.

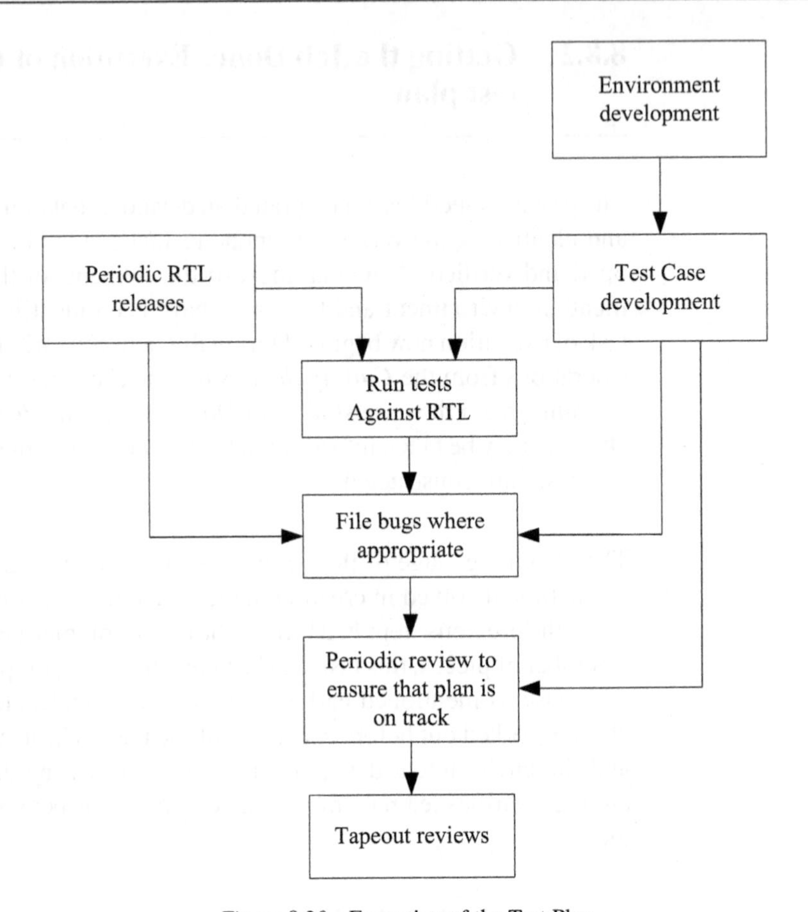

Figure 8.20. Execution of the Test Plan

On completion of the execution of the verification plan, the verification signoff review process begins. This process ensures that nothing got left out from the verification front as far as possible and that the device will work as intended.

8.8.3 Getting a Helping Hand from External Resources

Many a time, it becomes necessary for the verification team to get some external help via contractors or additional people in order to complete the verification on time. This happens frequently if there is some extreme risk to the schedule.

During such circumstances, it is imperative to specify exactly what is required of the person who is helping with the verification effort. It is also important that the person who is coming in to help the team is not a burden on the existing team's already stretched resources.

For instance: The example above in figure 8.1 now helps in clarifying the requirements to the person who will be helping the team. All that is required is to share the verification plan and request that the member to develop monitor *M3* or checker *cache2*. Such a specification is inbuilt into this process and hence there is no possibility of misunderstandings. (An example of habit 6: Communicate).

The other advantage of this process is that the person coming into the team is immediately productive. There is no need for the person to read all the specifications and understand what is to be done.

All that is required is that the implementation and implications of Monitor M3 or cache2 be understood. This is a much simpler problem to address since it becomes a straightforward implementation issue.

8.8.4 The Case for GATE Simulations

You do need to run at least a few simulations on a gate netlist

This topic has been a source of debate in recent years. Many a time, I have been presented with the argument that it is unnecessary to run a simulation on a GATE netlist. The author concedes that this may be true if the timing constraints used in the synthesis scripts are correct or if the design has a single clock of a single frequency. The gate simulations are designed to catch errors where the timing constraints are not what they are supposed to be. The designer definitely can provide input on the timing critical paths in the design. Appropriate care is definitely taken during the synthesis of these paths. However, it is impossible to mention ***all the other paths*** in the design, and these are the paths to be concerned about. Those paths are not the ones that are looked at in detail since there are quite a few of them.

Usually running a few Gate level simulations with a few well chosen tests is enough to verify the integrity and timing of the Gate netlist. Gate simulations are typically done at the end of the RTL development cycle. During this period, the Gate netlist is simulated with test cases to ensure that there is no problem with the design.

In some organizations, many a time, the Gate netlist is simulated with unit delays to flush out any problems with the netlist. The netlist is then annotated with SDF (standard delay format) files which provide post layout delay information across various process and temperature parameters.

There has been considerable discussion about eliminating Gate level simulations altogether. For single clock designs, static timing analysis and equivalence checks available are sometimes considered as alternatives for Gate level simulations. However, in many situations, it may not be possible to eliminate Gate simulations altogether.

Multicycle paths

Many a time, some transactions take multiple cycles to complete. Although these are described as multi-cycle paths during synthesis and timing analysis, It becomes essential to verify that the constraints were indeed interpreted correctly and the synthesized design performs as expected.

False paths through the design

Many a time, there are false paths through the design. Some of these may be accounted for in the synthesis constraints. However, there is always the possibility that something is missed out.

Clock domain crossings

Clock domain crossings are typical in many designs. Typically, care is taken during the synthesis phase as well as the RTL design phase. However, it is possible that not all the scenarios are accounted for. The RTL design is typically built using either unit level or no delays at all. Hence, it becomes challenging to uncover issues in a RTL level simulation. Gate level simulations will easily uncover such problems with the appropriate test cases.

Clock gating

Clock gating with the appropriate delays is another reason why Gate simulations are useful. Any clock that passes through a

Gate will be skewed with respect to a non Gated clock. There may be some situations where there might be problems because of the Gated clocks which can be uncovered in Gate simulations.

Tester data simulations

The Gate level simulations are usually used to generate Value Change Dumps (VCD) files which are then post processed to create files that can be used to qualify parts on the tester. The VCD files are usually converted into a TDL (tester description language) file used to control the tester. The tester uses the TDL files to test the device after it has been fabricated.

Power estimations

Gate level simulations after annotation can provide a good estimate of the power consumption of a device for some scenarios[8].

Verify the STA and synthesis constraints

One of the other important reasons to run a Gate level simulation is to verify the synthesis and Static timing analysis constraints. This verification is not about the constraints that have already been put in, but about the *ones that may have been missed.*

STA unfriendly designs

Many designs may have some STA unfriendly components in the designs like latches etc. Under these circumstances, it may be impossible to do an accurate timing analysis. Gate level simulations serve to increase the level of confidence of the verification engineer.

[8] It is noted that there are now several tools that perform this function. However, some organizations still use the approach of power estimation using a Gate simulation.

8.9 Step 9: Figuring out where you Are in the Process

In any project, verification is one of the few activities that takes the longest and according to some managers, costs too much, no matter what is done. Given the nature of the verification effort, it becomes imperative to ensure that all is going well with verification during the course of the project. Various metrics are available to gauge the effectiveness of the verification effort as a whole.
One of the common activities is to conduct a review where all aspects of the design are discussed with all the parties.

Why is the review process even important?

Is not a review process a hindrance in the first place? The author has been at a few places where the designers hated the review process. To them, it was a hindrance to their getting their job done. Some of the review conclusions were known to them anyway and to some was an absolute waste of time.

The review process is crucial. It can be kept as a short process, to save everyone time, but it must never be "short-circuited". The review process allows everyone to take a moment to get the bigger picture before they delve back into the detail of their activity.

Many of the organizations use some periodic review process. The nature and content of the review does vary, however, this procedure seems to be very useful overall to bring people not completely aware of the project details to understand the project's progress.

In this step, we present several methods that can be deployed successfully by any verification team to ensure that progress is indeed made on the design. Many of these methods rely on a weekly or bi-weekly schedule of activities. While the author does not particularly like to insist on a specific schedule, It is however emphasized that the review ***must*** follow some sort of a schedule which ensures that the review activity happens on a periodic basis.

Using tags to tell you where you are in the verification process

In the earlier steps, each test object was assigned a unique tag. This tag now comes to the rescue of the verification engineer. During the process of writing the test, it had been suggested that the test writer place the documentation in the same test file. A similar suggestion is also made to place the documentation for the checkers in the checker files.

A simple script can now build a correlation between the tests, the checkers and the monitors can now be deployed. This script will very easily reveal where the tests and monitors correlate well.
In the previous steps, various tests were written to test the compliance of the device with the specification. In this section, we discuss how we could tie these items all together.

In the test documentation portion of "cutting the ties that bind", a sample format for test documentation was discussed. It is possible to use some of these features now to bring together all the various components.

In the figure 8.21, we notice that there is an entry in the documentation section for a list of test objects that were covered by the test case. In the test documentation section, a list of test objects exercised by the test case is also listed. A structure similar to the figure indicated (possibly with some additional information captured would be used for the Monitors/checkers).

It hence becomes possible to create a "fact sheet" or a "status sheet" using the following algorithm shown in figure 8.22. Such a correlation is now possible since all the infrastructure was built when test development progressed. The correlation is now very easy having embedded the tags and other components in the test case when the test cases were developed.

The reader will note that it appears reasonably trivial! There is not a whole lot of engineering to it either!

The result of the correlation is now captured and could possibly be presented as shown in the figure 8.23. It is noted that tests that have been optimized to cover multiple objects will actually show up in the report for the other objects as well. For example: Cache2 and Cache4 could share a test. All the tests for a particular test object are now in the report. It now becomes

```
/*
*************************************************************************
Confidential information. Some Company. No rights without permission….
……….
….
*************************************************************************
TEST_DOCUMENTATION BEGIN

Test Name: fifo_register_overflow_1.v
Test Intent: Check the register for the fifo overflow bit.
Test Description:
This test checks the read and write of the fifo overflow
bit using a register write to an addressable register. Steps are:
1) Do a write to the RING_WRTR_ADDRESS register with the data for the bit
2) Read back the register along with the mask to make sure it compared correctly
3) Read the fifo register…..
…..
….


Test Assumptions: None
Test Notes: This test will run at both block and system level.
Test Results: Test is deemed pass if the register bit is set and no other errors are found.
Functional Objects Covered: fifo_tag_1, Fifo_reset_2, MMR_qualify_1,….

TEST_DOCUMENTATION END                          Use these tags to correlate!

*************************************************************************/
`include test,v

If(reg_write_main) begin
            // We have a register wrtite to the main block and not the decoder
             do_write(RING_WRTR_ADDRESS, RING_WRTR_MASK, data);
            do_check(RING_WRTR_ADDRESS, RING_WRTR_MASK, data);
….
…...

end //

…..
….
```

Figure 8.21. Using Test Documentation to Correlate

Step 1:
For all the test cases in the test case list
begin
For each test case, parse the test file and compile all the documentation information
End

Step 2:
For all the monitors and checkers in the list
begin
For each monitor, parse the code and compile all the documentation information
End

Step 3:
For each test object tag in the test object list
Begin
Get a list of monitors &checkers coded so far matching the test object tag.
Get a list of test cases coded so far matching the test object tag
Use the checker data structure from step 2 and the test data structure from step 3 to do the following:
Check them against the plan on the plan
Report any differences from the plan
End

Figure 8.22. Correlation Algorithm

possible to take a quick look and see whether there is anything missing from the testing of test object cache2. It also becomes very quickly evident if the tests for test object cache2 are the necessary and complete ones[9].

[9]The list of tests presented is obviously a partial list and for illustrative purposes only. The test object could itself be broken up into several sub objects like the ones for error testing only etc and the process repeated fairly trivially.

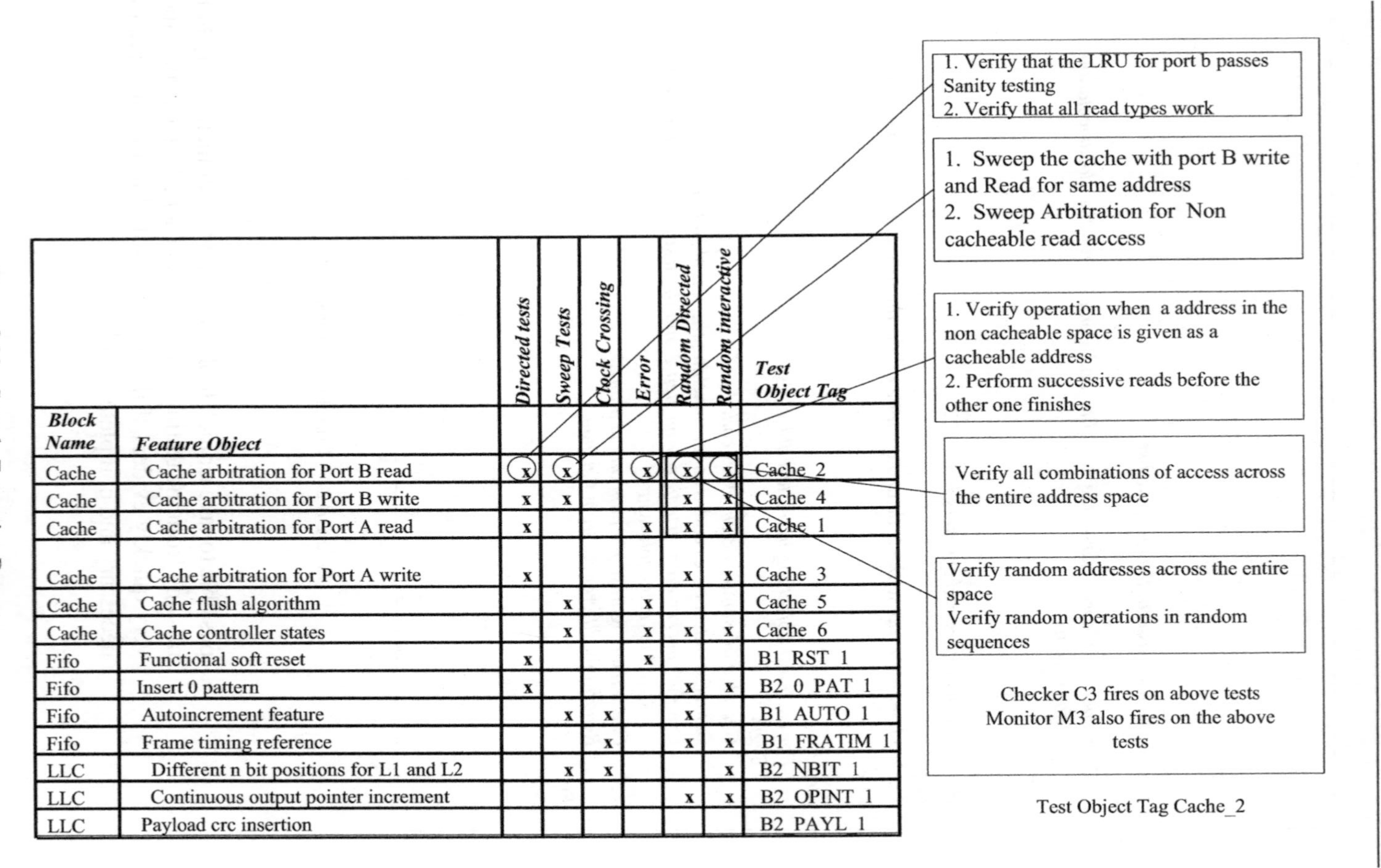

Block Name	Feature Object	Directed tests	Sweep Tests	Clock Crossing	Error	Random Directed	Random interactive	Test Object Tag
Cache	Cache arbitration for Port B read	x	x		x	x	x	Cache 2
Cache	Cache arbitration for Port B write	x	x			x	x	Cache 4
Cache	Cache arbitration for Port A read	x			x	x	x	Cache 1
Cache	Cache arbitration for Port A write	x				x	x	Cache 3
Cache	Cache flush algorithm		x		x			Cache 5
Cache	Cache controller states		x		x	x	x	Cache 6
Fifo	Functional soft reset	x			x			B1 RST 1
Fifo	Insert 0 pattern	x				x	x	B2 0 PAT 1
Fifo	Autoincrement feature		x	x		x		B1 AUTO 1
Fifo	Frame timing reference			x		x	x	B1 FRATIM 1
LLC	Different n bit positions for L1 and L2		x	x			x	B2 NBIT 1
LLC	Continuous output pointer increment					x	x	B2 OPINT 1
LLC	Payload crc insertion							B2 PAYL 1

1. Verify that the LRU for port b passes Sanity testing
2. Verify that all read types work
1. Sweep the cache with port B write and Read for same address
2. Sweep Arbitration for Non cacheable read access
1. Verify operation when a address in the non cacheable space is given as a cacheable address
2. Perform successive reads before the other one finishes
Verify all combinations of access across the entire address space
Verify random addresses across the entire space
Verify random operations in random sequences
Checker C3 fires on above tests
Monitor M3 also fires on the above tests
Test Object Tag Cache_2

Figure 8.23. Sample Extraction Report

Step 1:
For all the test cases in the list
Begin
For each testcase, parse the test file and compile all the documentation information
End

Step 2:
For all the monitors and checkers in the list
begin
For each monitor, parse the code and compile all the documentation information
End

Step 3:
For each test object tag in the test object list
Begin
Create an index of the test cases and monitors / checkers from the data structure above.
Print out a list of tests and the monitors for each tag
Identify the gaps using a correlation algorithm.
End

Figure 8.24. Correlation Using Tags

The tags from various files now help us bring together the various components of the verification environment so that we have a complete picture of the state of the design. During the process of writing all the tests that were identified in the test plan, the author suggests holding a few periodic reviews to ensure that everything is indeed going according to plan.

8.9.1 Performing Hole Analysis of What got Left Out in the Test Plan

During test development, it always happens that something or the other is left out and caught during a review. This is a normal occurrence. Doing the reviews periodically allows the reader to find problems sooner than later. That said, it is now possible to get a handle on the holes in the design using the steps in figure 8.24.

8.9.2 The (bi)Weekly Review Processes

While the author has indicated a review process on either a weekly or a biweekly basis, the intent of the process is to understand if appropriate and necessary thoroughness has been applied to each of the test objects that have been identified in the verification plan.

The principle behind this is to typically involve the designer and perhaps someone not working on the module who can offer a different perspective on the features tests.

The review process could be weekly/biweekly or whenever based on the team's opinion. It is defined to be something that is somewhat short. for every group of test objects. It is intended to go into low level details of the implementation. If the material has been made available to the reviewers beforehand (as can be facilitated using the algorithms given above) the authors opinion is that the review can go smoothly and quickly.

In my previous companies, good results have been obtained when the process is kept brief and spaced apart. In modules where due diligence was not followed, The price of a marathon session and a few complaints of "time wasted" seemed to abound. This seems to have usually been true for many of the modules though!

During the review, a quick code review of test code along with a feature mapping of the features tested against the plan can be undertaken. The bugs found and other possible scenarios that were missed out and need work can be discussed.

Conducting the review in this manner enables a few things: The test environment code is reviewed much like an RTL code review. It takes minimal time and not everyone needs to be present for the bi-weekly/weekly ones.

The other effect is that any bottlenecks or understanding of the implementation is ironed out quickly. The engineer who has been focused on the particular test objects for that period has already kept the test code fresh in his memory, hence the turnaround time to fix something will certainly be a lot less than if he was asked to fix something sometime later.

The review process *for each test object* could have the following questions asked:

1. Was the functional object understood correctly?
2. Did the test writer get the intent of the test? (i.e: Did we miss something?)
3. Are there any additional tests that were not specified on the plan?
4. If so, what was missed.? Did our understanding change?
5. Is there the need for any additional directed test cases?
6. Does the test need any additional random test cases based on coverage data?
7. Are all the negative tests conditions covered based on what is known during implementation?
8. If the functional object was at a higher level (not implementation friendly, was it broken down into some smaller feature objects?) did we cover all of them?
9. Is the testing complete? i.e. we don't need to test this test object any further?
10. If we chose random testing, did we get enough seeds into the simulation?

Any other questions may be added to this list or deleted from it as the reviewer feels fit. The basic intention is to ensure that a review process moves onward smoothly and covers the intent of verification

In addition, the reviewer can look at the test case densities report for some clues. Consider the example shown in figure 8.26. In the figure, one can see that the report for the correlations show that the test objects MAC_ARB1_STATE_START_1, MAC_ARB1_STATE_LOGIC_2 and others are actually covered in many tests. the feature MAIN_ARBIT_PRIORITY_1 is a feature which is tested only in the test case mac_eth_loop1 and not

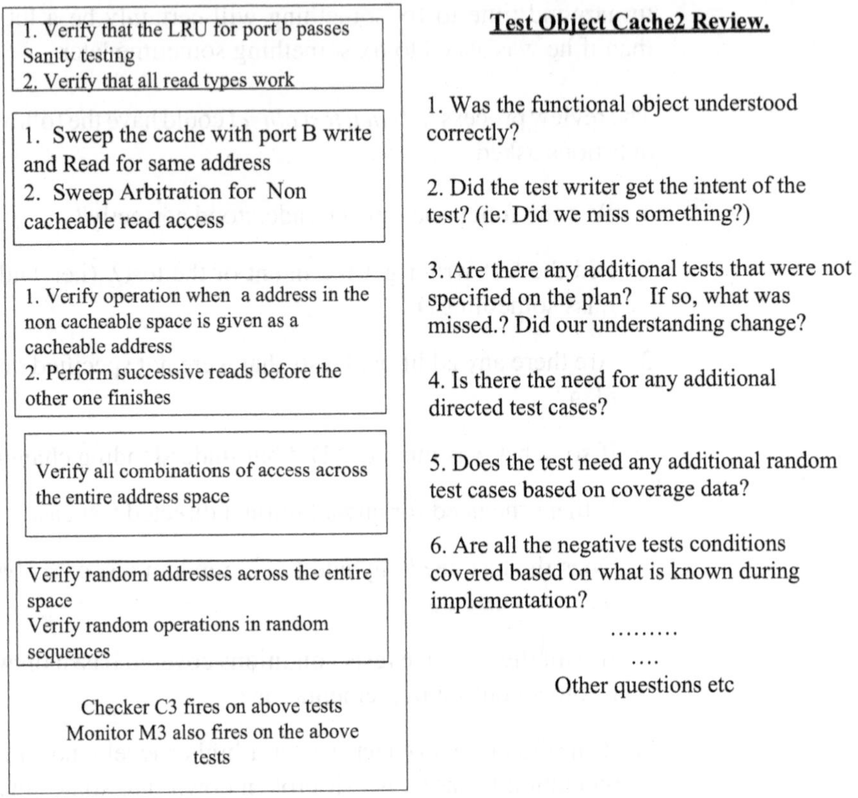

1. Verify that the LRU for port b passes Sanity testing
2. Verify that all read types work

1. Sweep the cache with port B write and Read for same address
2. Sweep Arbitration for Non cacheable read access

1. Verify operation when a address in the non cacheable space is given as a cacheable address
2. Perform successive reads before the other one finishes

Verify all combinations of access across the entire address space

Verify random addresses across the entire space
Verify random operations in random sequences

Checker C3 fires on above tests
Monitor M3 also fires on the above tests

Test Object Cache2 Review.

1. Was the functional object understood correctly?

2. Did the test writer get the intent of the test? (ie: Did we miss something?)

3. Are there any additional tests that were not specified on the plan? If so, what was missed.? Did our understanding change?

4. Is there the need for any additional directed test cases?

5. Does the test need any additional random test cases based on coverage data?

6. Are all the negative tests conditions covered based on what is known during implementation?

....

....

Other questions etc

Figure 8.25. Reviews For a Test Object

tested anywhere else in other test cases. Hence it becomes imperative that there be a review to ensure that the feature MAIN_ARBIT_PRIORITY_1 is tested for all possible scenarios in the mac_eth_loop1 test case. Otherwise there is a possibility that there are some scenarios for testing which have escaped the verification activity.

Change does happen even after all this work is done. We all live in a world of change. However, once a object is deemed completely verified, there is usually not much push to "get rid of it" unless there is some pressing area or market pressures. Any change to the test object is usually deemed easy to determine and complete due to the above review. If a feature is deleted, the elimination or modification of affected tests is now

Test Case Density report

List of test cases for each test Object -- Click on test cases to see descriptions of each test, The number of links under each test object indicate the number of tests for that test object.

MAC_ARB1_STATE_START_1

mac_arbiter_full_chip_priority1
mac_arbiter_full_chip_priority_with_boot
mac_arbiter_sub_chip_priority_with_ring
mac_arbiter_check_all_states
mac_eth_loop_1

MAC_ARB1_STATE_LOGIC_2

mac_arbiter_full_chip_priority1
mac_arbiter_check_all_states
mac_eth_loop_1

MAIN_ARBIT_PRIORITY_1

mac_eth_loop_1

MAC_PKT_TRANSMIT_START_FINISH

mac_eth_loop_1
mac_arbiter_sub_chip_priority_with_ring
mac_arbiter_full_chip_priority_with_boot

Test Object Density Report

This report is an extract from the test cases that reveal the number of test objects tested by that test case.

mac_arbiter_full_chip_priority1:

MAC_ARB1_STATE_START_1,
MAC_FIFO_STATE_START_1,

mac_eth_loop_1:

MAIN_ARBIT_PRIORITY_1,
MAC_ARB1_STATE_START_1
MAC_ARB1_STATE_START_2
MAC_PKT_TRANSMIT_START_FINISH
MAC_PKT_RECEIVE_START_FINISH
MAC_FULL_LOOP_TEST,MAC_SELF_TEST
MAC_ARB1_STATE_LOGIC_2

mac_arbiter_full_chip_priority_with_boot:

MAC_ARB1_STATE_START_1
MAC_PKT_RECEIVE_START_FINISH
MAC_FULL_LOOP_TEST,MAC_SELF_TEST
MAC_PKT_TRANSMIT_START_FINISH

mac_arbiter_sub_chip_priority_with_ring:

MAC_PKT_TRANSMIT_START_FINISH

Figure 8.26. Test Case Density

a straightforward task.

Everything is now a bite sized chunk in the verification plan

What has effectively been done is to break up the entire verification plan into a series of very small chunks. Each of the small chunks could be reviewed for completeness and closed on a periodic basis. The reviewer of the code could possibly be another person other than the one who developed the code.

The other side effect of this is that a script could pretty much print out only the relevant pieces of code into a single file so that a reviewer could peruse it. Since the reviewer is not looking at the entire verification code base, the reviewer's job gets a whole lot easier since all that is accomplished is review a very small collection of files related only to that test object and offer feedback. This cuts down significantly on the workload of members of a team.

Just pay attention to the exceptions!

For each test object, it now becomes a trivial task to extract the exceptions to the above rules or others that may be agreed upon by the verification team. Any test objects which reveal a status not the norm are the only ones a verification team has to look at.

The representation of the tests in the above plan in some form ensures that most code coverage metrics are at their desired targets automatically. Any code that does not have coverage at the stated goal can have the following implications:

1. The RTL code is possibly dead code.
2. The verification plan is not complete – Something is missing from the test object list or the test case list.
3. Something is wrong with the test implementation.
4. This should have been caught in either the reviews done periodically.
5. Something is constrained in a random environment.

The dynamic coverage that is obtained from simulation is also included in the review process. This is shown on the right hand flow in the figure 8.27.

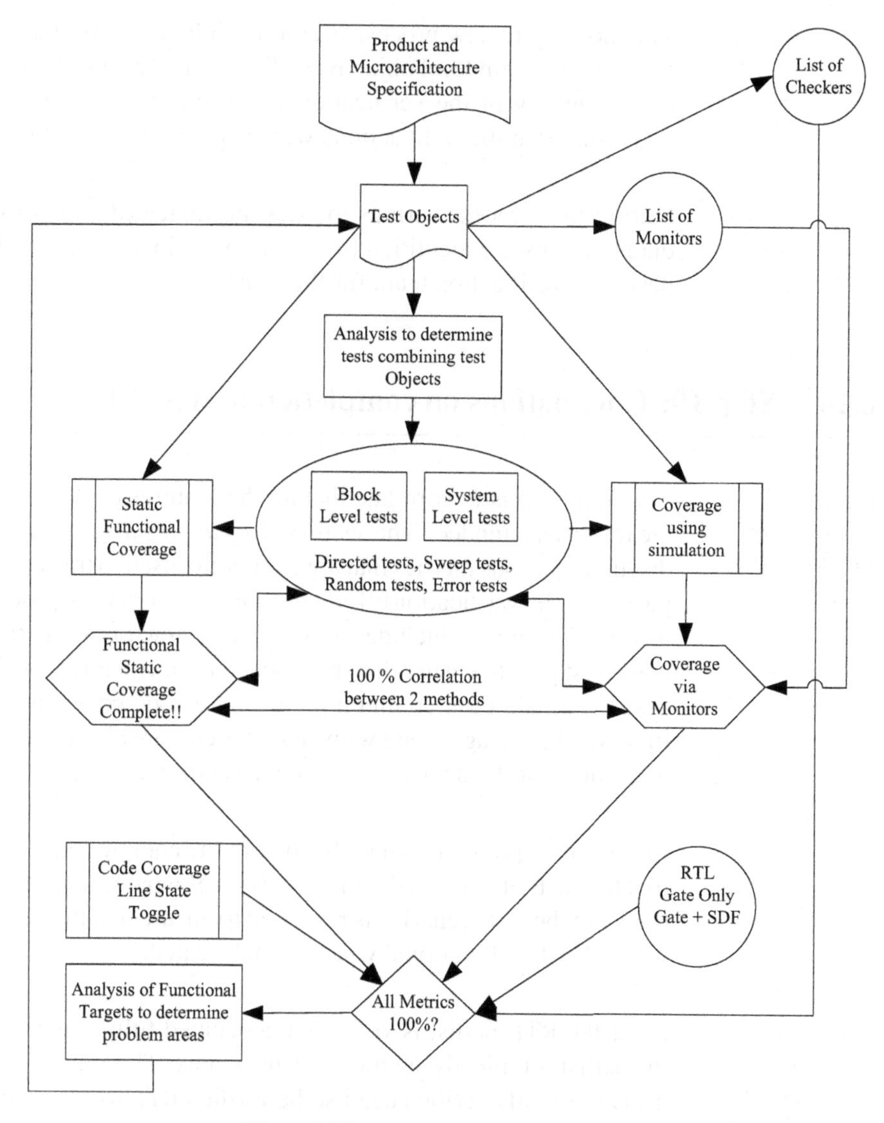

Figure 8.27. Complete Correlation Flow

8.9.3 The monthly Review Processes

The monthly review is seen in many organizations. It is a process that allows the team to take a step back and look at what got accomplished and identify the big picture moving forward in the verification effort.

The monthly review process is at a much higher level that focuses on program level objectives. The earlier reviews focused on the quality of the verification effort and were targeted to make sure that the right actions were happening all the time.

During this process various metrics are reviewed and consequent actions are identified that need to be done by either the design or verification team for the project.

8.10 Step 10: Correlations on completion to sign-off

Ensuring completion of various metrics.

The final step is to correlate that all the features of the device are tested as intended. The section *Tracking results that matter* discusses various metrics that are commonly used to track completion of the verification effort. In addition, the review process to signoff typically includes a thorough review of all the tests in the test plan to ensure that they meet the specifications of the device. In addition, common metrics like code coverage and functional coverage along with bug rates etc. are also reviewed to ensure that the device is indeed ready for tape-out.

In case the tape-out is scheduled before the completion of verification activity, then a detailed analysis of what has been verified and the test scenarios is reviewed to ensure that the device will at least be functional when it is fabricated.

Running it on various Process and temperature, voltage corners

After the RTL netlist is verified and deemed fully functional, the netlist is typically synthesized into Gates. The Gates netlist is also formally verified against the verified RTL to ensure that the Gate netlist is an accurate and true representation of the RTL netlist.

The ASIC vendor typically provides library data for the various process corners. This information to back annotate the netlist is typically made available shortly after the design is routed in the physical design phase. It is recommended that a few simulations be run with this information so that the reader can ensure that the circuits will behave as intended over various process corners.

Conclusions

A template is built from the ground up

The previous ten steps outlined a 10 step verification plan that is comprehensive and complete. From a starting point where there was no verification plan, a full verification plan was built. Various checkers and monitors were identified as a result of this process and a chart created has allowed the reader to measure tangible progress.

The necessary and complete criteria are built in!

Almost all verification efforts suffer from a paucity of time to get the job done. Consequently, the question that is asked is whether the necessary tests are written and if the verification plan is complete.

If each of the features are completely understood and the corresponding tests and monitors are developed according to the verification plan, then it can be shown that the plan that had been built in the above steps has the information to declare that the *necessary and complete* tests have indeed been a part of the verification plan.

The periodic reviews suggested reinforce the fact that every engineer is indeed on the same page as far as test development is concerned.

Ability to "Bridge" to formal verification methods

At some point, it may be possible to formally verify the entire design. Analysis of test objects results in a list of checkers and monitors which can then be conveniently translated into properties in formal verification. The translation effectively becomes a format conversion if the approach is defined correctly.

The reader may choose to implement some part of the verification plan using a formal based approach and others using simulation. This approach is widely demonstrated by the offering of tools by many vendors like Cadence, Synopsys, Mentor and others.

Quality is built right into the process

The reader will note that there is a check-and-balance system that is built into the process. In typical verification environments, the checkers and the tests are built by different individuals. The tests check the checkers and monitors and vice-versa. In addition, the process of tagging items ensures that the relationships which were conceived because of several peer reviews are now captured as intent of the tests. Most of the tracking aspects of verification are now an inbuilt part of the process in addition to better predictability.

Minimized reporting!

Many readers may wonder *"isn't this a lot of data to keep track of?"*. *"We did not do many of these steps and still got a working chip!"* Or *"I don't think I need any of this!"*

One of the key differences the author would like to point is that when building any type of SOC or other device where verification IP is being heavily reused, the challenge is more in keeping track of things than the actual work itself. Many a time, the data that was collected was collected using some straightforward PERL scripts and neatly formatted so that the reporting was more or less automatic without the intervention of the engineer. The engineer only had to share what was covered and what issues he faced during a quick review and that was it! The rest of the story is easily told using a script.

Hierarchical Encapsulation of the problem

The approach described above is a "divide and conquer" approach. Each test Object is encapsulated into an object oriented approach wherein the scope of the problem is reduced dramatically. The effect is that it allows the engineers to think in "small chunks". The author has been approached many a time by junior verification engineers who just ask me the question: *"Tell me what you want me to do next so that I can help!"*

The other side effect of this approach is that it is possible to give different people different aspects of the verification challenge being secure in the knowledge that it will all come together to form a successful package without errors and redundancy. One engineer can develop the checkers and/or monitors and another can focus on the tests.

Based on the relationships we had built above, it becomes possible to ensure that the quality of the verification environment is very high. The tests exercise the design and the checkers and monitors and vice versa. Since different people are contributing, an automatic "*check and balance*" is automatically achieved.

Schedule predictability is inherent in this process

As can be observed, unless the features change or an understanding of the feature changes dramatically, the schedule predictions as part of the process are usually close to accurate.

Ability to see the big picture throughout the project

The focus on test objects allows the test engineer to identify the features required in the environment up front. The engineer is able to stay focused on the item being worked on.

At any point of time, a few standard reports that have been agreed to by the team can be used to determine if the features are tested or not. The review process is designed to reveal any holes in the methodology and the ability of the team to complete the tests.

Ability to get "Real time" Status on the verification effort

The verification effort now has tags embedded in the tests, the monitors and the checkers. Hence, it now becomes easily possible to use some of the simple algorithms outlined above to get a clear cut status of the verification effort.

It is possible at any point of time to use a simple method of looking for the test object tags in the tests and the monitors/checkers to present on a verification status.

Ability to spec out the problem for Outsourcing/. "Rightsourcing?"

One of the biggest challenges for outsourcing any portion of the design is that the specification is a "soft" specification. Understanding and communicating ideas is also a challenge. A clear specification enables the completion of a goal on time and on schedule.

The earlier steps revealed how each test, checker and monitor are now explicitly specified. If outsourcing of a certain number of tests/checkers/monitors is planned on, then the complete specification is available directly from the Test Graph Matrix.

Consequently, it will be possible to come up with a list of test objects that can effectively given to another team for execution. Measurement of deliverables becomes possible since the coverage (both static and dynamic) is measured using a set of standard tools that are available today.

In short, the benefits of the 10 step methodology presented above are many. The author is confident that the concepts presented above are practiced in some shape or form by almost all the organizations in the ASIC industry. It is hoped that the above gives the reader a good insight into the verification effort and allows the engineer to be successful in their careers.

References and Additional reading

[1] Kuhn, T., Oppold, T., Winterholer, M., Rosenstiel, W., Edwards, Marc, and Kashai, Yaron (2001b). A framework for object oriented hardware specification, verification, and synthesis. In *DAC '01: Proceedings of the 38th conference on Design automation*, pages 413–418, New York, NY, USA. ACM Press.

[2] Bergeron, Janick (2003?). *Writing testbenches - functional verification of HDL models.* Kluwer Academic Publishers, Boston, 2nd ed edition.

[3] Foster, Harry, Krolnik, Adam, and Lacey, David (c2003). *Assertion-based design.* Kluwer Academic, Boston, MA.

[4] Meyer, Andreas (2004). *Principles of functional verification.* Newnes, Amsterdam.

[5] Peet, James (2001). *Template for a verification plan.* www.everaconsulting.org.

[6] Peet, James (2000). *The five day verification plan.* www.everaconsulting.org.

[7] Albin, Ken (2001). Nuts and bolts of core and SoC verification. In *DAC '01: Proceedings of the 38th conference on Design automation*, pages 249–252, New York, NY, USA. ACM Press.

[8] Synopsys Inc(2005). Hybrid formal verification.

[9] Yee, Steve (2004). *Best Practices for a Reusable Verification Environment.* www.design-reuse.com, New York, NY, USA.

GLOSSARY

SOC	System On a Chip
ASIC	Application Specific Integrated Circuit
BFM	Bus Functional Model
HDL	Hardware Description Language
RTL	Register Transfer Language
IP	Intellectual Property
ABV	Assertion Based Verification
DFV	Dynamic Formal Verification
DDFV	Deep Dynamic Formal Verification
DUT	Device Under Test
PCI	Peripheral Component Interconnect
HDL	Hardware Description Language
HVL	Hardware Verification Language
IP	Intellectual Property
PLI	Programming Language Interface
SDF	Standard Delay Format
PERL	Practical Extraction and Reporting Language
TCL	Tool Command Language
FPGA	Field Programmable Gate Array

Appendix A
Using PERL to connect to Microsoft Excel and Access

This appendix provides information that has been put together by the author to help readers to set up a database like mechanism on their windows desktop computers so that they may be able to generate data and reports easily. The main motivation for this appendix came from the fact that most verification engineers are typically UNIX/Linux users and many a time, the expertise on Microsoft programs is not usually readily available. In many organizations, the verification work is done on UNIX like machines while the presentations and other spreadsheets have been maintained typically as Microsoft Access or Microsoft Excel programs[1].

This appendix shows the reader step by step instructions to hook up PERL to a Microsoft spreadsheet or Microsoft database so that SQL like commands may be deployed and the full power of PERL is available. The best of Microsoft programs and PERL will then be available to the user. Detailed screenshots to help users through the process will also be available on http://www.effective-verification.com/ along with other similar techniques.

This information is culled from many sources. There are no doubt many sources on the internet which provide details of each of the steps presented herein. Readers wishing to learn more are requested to search for this information using their favorite internet search engine[2].

---

[1] Many products are mentioned here. The trademarks belong to the respective companies

[2] The information presented here is deemed to be accurate at the time of writing. It has been used by the author effectively in many situations and provided to assist fellow verification engineers. However, neither the publisher nor the author can assume any responsibility for the fitness of these instructions for any purpose. Please use them as you will at your own risk.

Note: The commands that the user has to type in is presented in *italics*.

Step 1:
Ensure you have administrative rights and install Activestate Perl. This release of Perl is extremely popular among Windows users. It is available from http://www.activestate.com/.
If you use the regular PERL installation, you will be able to use the cpan> shell instead of the ppm environment described below.

Step 2:
After the above program is installed,in your windows environment, Click Start -> Run
In the input box type in the command *cmd*

A Dos window will open on the screen.
Step 3:
If you use a proxy to access the internet, type in the command
set HTTP_PROXY=<whatever is the proxy> (Do not use the angular braces)

1. Type in the command *ppm* The reader will notice a *ppm>* prompt.
2. At the prompt, type *install dbi*

   The Perl Package Manager (ppm) should go off and download some material from the internet if necessary. It should return the **ppm**> prompt when it is completed.
3. At the prompt, type *install dbd-odbc*
   Many drivers will get installed on your computer. These drivers help you to access the database easily.
4. Type exit to exit the ppm shell
5. Type exit to exit the dos command window.

Step 4:
The hardest part is complete. The next step is to set up an source/destination to which the PERL interface has to connect to. This is typically the file that the reader would like to access in PERL. The reader will have to repeat steps 1-8 for each EXCEL/Access file that needs to be accessed.

1. Click Start ->Control Panel -> Administrative Tools -> ODBC
2. Start up the ODBC applet by double clicking on it.
3. Click on Add

   (a). Choose the type of file. There are many choices from them. for Excel Files, you may choose the Microsoft Excel Driver.

   (b). Click Finish

4. You will see another window for ODBC data sources. In this window, Create a Data Source name. This should NOT contain any spaces.

```
## This example is written by the author based on his prior experience.
## Search the web for "PERL ODBC example" to get more detailed
examples using
## your favorite search engine and you will see similar information

#!/usr/bin/perl
use DBI;
use Data::Dumper;

## the name in the next line must be replaced with the ODBC source you
created.
my $dbh = DBI->connect("dbi:ODBC:<name>","root","",)
        or die "Unable to connect:".$DBI::errstr."\n";
# set the length of the
        $dbh->{LongReadLen} = 100000;
        # This is to clear out the existing stuff

## tablename is the name of the table/sheet in the ODBC setup.

 my $sql = "SELECT * FROM tablename";
 my $sth = $dbh->prepare($sql) or die "preparing: ",$dbh->errstr;
 $sth->execute();

                while ($row = $sth->fetchrow_hashref)
                {
                        push(@result,$row);
                }
                $sth->finish;

## Now @result is a PERL array that has all the results from the database.
## You can use other SQL commands similar to the above example
```

5. Type in a description in the description field.
6. Click on Select Workbook and choose the file that is required.
7. Click OK
8. You should now see the Data Source in the list of data sources on your machine

Step 5:
Connect to the database from the PERL script. Sample code is shown below.

Appendix B
Using PERL to convert between UNIX text files and Microsoft Word

The chapter *Cutting the Ties that Bind* presented an Automatic Documentation methodology which allowed users to significantly reduce the work that they had to do when generating documentation. The information presented below is a collection of basic HTML techniques which have been used by the author to implement the concept presented.

Assuming that the test documentation is placed in the test, the reader will notice that there are some times when a complete verification test description document is needed with all the tests developed so far. Since the information now resides in many files, a simple parser script can be used to get this information into a data structure in PERL. This information can then be neatly formatted automatically and saved as a single document periodically.

It is acknowledged that there are many ways to achieve the above result. What is presented is only *one of very many* ways. The reader is encouraged to use this information to implement a solution that fits the readers own unique situation. The following is intended as a general guide to help the reader get started.

Step 1:
Create a document in the favorite WYSIWYG (What you see is what you get) editor of your choice. This editor must be capable of generating and reading in HTML. You can add all the graphics and other material you need to make the document as per your needs.

Step 2:

1. In the document, identify the location where the test descriptions have to reside.
2. Mark this location in the document with a START_HERE marker. (Change the text to your liking, it only has to be unique.)
3. Create a dummy test description with some arbitary data with the style (Font, numbering etc) that is appropriate for the situation.
4. At the end of this test description, put in a END_HERE or similar marker.
5. Save this file in its native format and HTML format.
6. In the steps that follow, the HTML file in step 5 is called a template file. The author recommends that the reader save it in a separate directory since the graphics may render into some sort of images.

Step 3:

1. It is assumed that the reader has implemented a simple parser to read in information from the various test files. A template for the styles for the test descriptions was created in the previous step.
2. Open the file using a text editor as a text file. Cut out everything between the START_HERE and END_HERE into a second file. This serves as your test template for each of the descriptions.
3. Save the template HTML file.
4. Use the perl script to open the Template file and reach the START_HERE point in the file. Print all the contents of this template file into a output HTML file upto the START_HERE point. (as shown in pseudo code below).
5. For each test, format the test as per the test template and print this information into the second HTML file. (shown in print_test_descriptions pseudo code below).
6. Copy anything after the END_HERE from the template file into the output file.
7. Close all the files. Open the HTML file either with your browser or favorite editor and print/save it as required.

An Example:
Assuming that the template file is called template.htm, the test description is called test_desc.htm,
The template file would have the following
<HTML>
....
some html text
.... START_HERE HTML describing the test description
....
END_HERE
...
...

Some more HTML . . . </HTML>
The test description template file that is created Could have the following (everything between START_HERE and END_HERE tags after omitting the tags: <b>Test Name: </b> name of the test. <b> Test Intent: </b> Intent of the test <b> Test Description> Test Description ..
any other items from the test that are required as per test template. It will be obvious to the reader as they carry out these instructions.

A sample pseudo code in PERL is presented in the example below. The code is not complete and is intended only as an illustration. The reader can easily create something based on the template. and the details are left as an exercise to the reader.

```
#!/usr/bin/perl –w
## user variables.
my $test_template_file = "";
my $test_description_template_file = "";
my $output_file = $ARGV[1];
## open files.
open(TEMPLATE_FILE,"$test_template_file") || die "Cannot open test template file\n";
open(OUTPUT_FILE,"$output_file") || die "Cannot open output file\n";
## get the test descriptions
&get_test_descriptions;
## copy everything from TEMPLATE_FILE to the output file
$state = 0;

while(<TEMPLATE_FILE>) {
            if(/START_HERE/) {
                        $state = 1;
                        ## Print the test descriptions
                        &print_test_descriptions;
            }
            if(/END_HERE/) {
                        $state = 0;
                        next;
            }
            if($state == 0)
                         print OUTPUT_FILE $_;
            }
} ## while
close(TEMPLATE_FILE);
close(OUTPUT_FILE);

sub get_test_descriptions {
## Implement your test case parser here.
## Details not provided as they vary from situation to situation

## It is assumed that the test descriptions are in some sort of a  array of hashes
}

sub print_test_descriptions {
my $i;
$number_descriptions = $#test_desc_hash_array;
   for($i = 0; $i <= $number_descriptions; $i++) {
            ## create some local variables
            # $test_name
            # $test_description
            print OUTPUT_FILE "
            <b> Test Name: </b> $test_name
            <b> Test Description <b> $test_description
            ...
            ";
   } # end for
} # End sub
```

Figure B.1. Sample Code for Automatic documentation

About the Author

Srivatsa Vasudevan has spent much of his working career working in the field of ASIC verification. He has worked in various companies as a verification engineer in Silicon Valley, USA and Bangalore, India. He has worked on a variety of designs ranging from microprocessors to networking, wireless and multimedia devices. Much of his work involved developing testbenches and tests to verify devices in addition to developing methodologies. His interests include verification methodology and techniques to enhance time to market while delivering successful products. He has worked in many companies like Philips, Fujitsu, Crimson Microsystems, Ciena and Texas Instruments during the course of his career and has had the opportunity to work with and interact with many engineers in many parts of the globe. The author was formerly a Member of Group Technical Staff at Texas Instruments. He is currently Engineer, Senior Staff/Manager at Qualcomm Inc.

Index